RESOURCES FOR THE FUTURE LIBRARY COLLECTION
ENERGY POLICY

Volume 4

Economic Aspects of Oil Conservation Regulation

Full list of titles in the set
ENERGY POLICY

Economic Aspects of Oil Conservation Regulation

Wallace F. Lovejoy and Paul T. Homan

RESOURCES FOR THE FUTURE

Washington, DC • London

First published in 1967 by The Johns Hopkins University Press for Resources for the Future

This edition first published in 2011 by RFF Press, an imprint of Earthscan

Earthscan LLC, 1616 P Street, NW, Washington, DC 20036, USA
Earthscan Ltd, Dunstan House, 14a St Cross Street, London EC1N 8XA, UK
Earthscan publishes in association with the International Institute for Environment and Development

For more information on RFF Press and Earthscan publications, see www. rffpress.org and www.earthscan.co.uk or write to earthinfo@earthscan.co.uk

ISBN: 978-1-61726-018-6 (Volume 4)
ISBN: 978-1-61726-002-5 (Energy Policy set)
ISBN: 978-1-61726-000-1 (Resources for the Future Library Collection)

A catalogue record for this book is available from the British Library

Publisher's note

The publisher has made every effort to ensure the quality of this reprint, but points out that some imperfections in the original copies may be apparent.

At Earthscan we strive to minimize our environmental impacts and carbon footprint through reducing waste, recycling and offsetting our CO_2 emissions, including those created through publication of this book. For more details of our environmental policy, see www.earthscan.co.uk.

ECONOMIC ASPECTS OF OIL CONSERVATION REGULATION

Economic Aspects of

OIL CONSERVATION REGULATION

by

Wallace F. Lovejoy and Paul T. Homan

*Published for Resources for the Future, Inc.
by The Johns Hopkins Press, Baltimore*

This book is one of RFF's studies in energy and mineral resources, which are
directed by Sam H. Schurr. The research was supported by a grant to Southern
Methodist University. Wallace F. Lovejoy is professor of economics at the
University, and Paul T. Homan is consultant with RFF and former director of
graduate studies, Department of Economics at the University. The manuscript
was edited by Elizabeth Reed.

Director of RFF publications, Henry Jarrett; *editor,* Vera W. Dodds;
associate editor, Nora E. Roots.

Foreword

The production of crude oil is subject to detailed regulation by
agencies of most of the important oil-producing states. These systems of state regulation were established to prevent the glaring
wastes in oil production and disastrously unstable market conditions that had been plaguing the industry. Yet, today, regulation is
criticized on the ground that it encourages prodigious waste through
inefficient development and production of the country's oil
resources.

It is not surprising that the regulation of an industry should produce unintended consequences, particularly with the passage of time
and the failure of regulatory practices to keep pace with the industry's technical and economic evolution and with changes in the
competitive setting within which it operates. Factors such as these
would offer sufficient reason to scrutinize the present state of regulatory processes, in order to see how well they may be serving the
interests of the domestic economy.

There is also another reason. While the regulatory system operates at the level of individual oil-producing states, it exerts a pro-
found influence on the world position of the United States. During
the Suez crisis, for example, the ability of the United States to help
meet the emergency oil needs of Western Europe was found to depend upon decisions of the Texas Railroad Commission, the oil
regulatory body in that state.

Oil is regarded as a strategic material from the standpoint of
national security. It is also an internationally-traded commodity of
the first rank. The price the nation pays in order to maintain a

v

domestic oil industry that supplies a large fraction of domestic consumption (a condition that is deemed essential for security reasons)
is obviously affected by comparative oil production costs in the
United States and other producing regions of the world. The latter
relationship is, in turn, strongly affected by state regulatory practices in this country. By the same token, the nature of the trade
barriers that the United States has erected to protect domestic production is affected, as are the commercial relationships between the
United States and oil exporting countries.

In sum, state regulation of oil production practices critically
affects the performance of the domestic arm of an industry which is
strategic in nature and international in scope; and therein lie the
roots of a number of critical national policy concerns.

It is, therefore, important to understand why conservation regulation arose in the first place, the manner in which the regulations
have been carried out in practice, the results that have been produced and the reasons for them, and the possibilities for improvement contained in such remedial actions as have been proposed.
This is the task that Wallace Lovejoy and Paul Homan have undertaken in the present study. They have, I believe, made their way
through complicated and controversial subject matter with skill,
care, and objectivity, and have reported their results with admirable
clarity.

SAM H. SCHURR
Director, Energy and
Mineral Resources Program
January, 1967   Resources for the Future, Inc.

Preface

The present study originated as a background paper distributed in advance to the participants in a three-day seminar-conference convened by the Department of Economics at Southern Methodist University under a grant from Resources for the Future, Inc. The participants included representatives of oil companies, large and small, representatives of state regulatory agencies, university economists and lawyers, and independent experts. The common factor among them was an intimate knowledge of state regulation of the oil industry. The purpose of the seminar was to provide an exchange of views among persons who saw the problems of regulation from different angles and to stimulate thinking about the means to constructive improvement in the regulatory process and in the economic performance of the industry. The background paper was designed to provide the point of departure for such a dialogue. In order to encourage uninhibited discussion, no transcript was made. No vote was taken or consensus attempted on any question. We have unfortunately found it impossible to summarize the discussion, which developed in many directions with unsystematic variety. All we can offer, therefore, is our own analysis without an account of varied reactions to it.

The present study retains in large part the organization and tenor of the original paper. Specific parts have been revised extensively in the light of the discussion and of the written comments submitted by seminar participants and by other individuals connected with or interested in the oil industry and its regulatory agencies. Although economic in character, the study is relatively non-technical in language, being aimed not at an audience of professional economists, but at the broad spectrum of individuals and agencies interested in the economic aspects of oil conservation regulation.

Since conservation statutes and regulations help to shape the framework within which investment decisions are made by the domestic producing industry, such regulations are major determinants of the costs of finding, developing, and producing oil in the United States. Regulations and statutes can force efficiency or inefficiency on the industry. Some of the regulatory patterns today tend to foster inefficiency in an industry which can ill-afford it. The causes and aspects of this inefficiency are numerous and complex, often the result of the imposition of regulations in the past when the technical knowledge of oil production was highly imperfect, and much affected by conflicting property interests. The paths to correct the inefficiencies are cluttered with a myriad institutional, legal, and political hurdles. We have found no panacea. We hope that we have clarified the nature of the problems and identified the regulatory means available for improving the economic performance of the industry.

Lest some readers find this study to be all criticism and little praise, we must state at the outset that the existing regulatory system has accomplished a great deal in preventing the kinds of waste and instability that once occurred under an unfettered system. Looking back with today's knowledge it is easy to see what might have been done differently to advantage. We are not particularly concerned with criticizing past actions. We are concerned with the present economic status of the industry and the system of incentives and disincentives partly created by the system of regulation. The future health and vitality of the domestic oil industry is, we judge, a matter of importance in the public interest, as well as in the interest of members of the industry faced with foreign and domestic sources of competing energy supplies.

It is impossible to name all the people who have contributed their time and thought in commenting on earlier drafts of this study. The seminar participants each played an important part in this final product. In particular we wish to thank Sam H. Schurr of Resources for the Future, Inc., for his guidance and encouragement throughout this project. Also, we are deeply indebted to Robert E. Hardwicke who on many occasions engaged us in friendly debate and forced us to clarify or abandon positions we had taken. In addition to seminar participants, we would like to thank the following individuals and groups for their assistance: Joseph Lerner, Office of Emergency Planning, Executive Office of the President; the Office of Oil and Gas, U.S. Department of the Interior; the staff of the Independent Petroleum Association of America; the staff of the Texas Independent Producers and Royalty Owners Association; Granville Dutton, Sun Oil Company; Professor A. E. Kahn, Cornell University; Professor Joel B. Dirlam, University of Rhode Island; Professor R. F. Rooney, University of California, Los Angeles; Bruce C. Netschert, National Economic Research Associates; the Interstate Oil Compact Commission; the Texas Railroad Commission; the Kansas Corporation Commission; the Oklahoma Corporation Commission; and the Louisiana Department of Conservation.

Finally, we wish to thank Resources for the Future, Inc., for the grant which made this study possible.

The frame of reference, the analysis, and the conclusions in the study are, of course, the responsibility of the authors and do not reflect the views of the individual members of the seminar, the seminar as a group, or the other individuals or groups making comments on earlier drafts.

Participants in the seminar are listed on page x.

WALLACE F. LOVEJOY

August, 1966   PAUL T. HOMAN

Participants in the Southern Methodist University Seminar

(Affiliation of participants given as of the date of the seminar, April 1964.)

M. A. Adelman
  Department of Economics,
  Massachusetts Institute of
  Technology
Edward A. Albares
  Director, Arkansas Oil and Gas
  Commission
Lawrence R. Alley
  Executive Secretary, Interstate
  Oil Compact Commission
A. R. Ballou
  Proration Section, Sun Oil
  Company
Richard C. Byrd
  Formerly Chairman, Kansas
  Corporation Commission, and
  Administrative Assistant to
  Interstate Oil Compact
  Commission, Governors' Special
  Study Committee to Study the
  Conservation of Oil and Gas
Melvin G. deChazeau
  Department of Economics,
  Cornell University
Northcutt Ely
  Attorney and author in oil, gas,
  and mineral matters
William J. Flittie
  Law School, Southern Methodist
  University
Charles O. Galvin
  Dean, Law School, Southern
  Methodist University
Robert E. Hardwicke
  Attorney and author in oil and
  gas conservation matters
S. F. Holmesly
  Coordinator, Conservation and

Regulation, Humble Oil and
  Refining Company
Paul T. Homan
  Department of Economics,
  Southern Methodist University
Wallace F. Lovejoy
  Department of Economics,
  Southern Methodist University
Stephen L. McDonald
  Department of Economics, The
  University of Texas
William J. Murray, Jr.
  Formerly Commissioner,
  Railroad Commission of Texas
James E. Russell
  Engineering consultant and
  independent oil producer,
  formerly Chairman, Oil Industry
  Advisory Committee to Railroad
  Commission of Texas
Sam H. Schurr
  Resources for the Future, Inc.
E. Bruce Street
  Independent oil producer and
  formerly President, Texas
  Independent Producers and
  Royalty Owners Association
Thomas M. Winfiele
  Chief Engineer, Louisiana
  Department of Conservation
Foley F. Wright
  Manager, Engineering
  Department, Sinclair Oil and
  Gas Company
Fred Young
  Legal Counsel, Oil and Gas
  Division, Railroad Commission
  of Texas

Contents

List of Tables

ECONOMIC ASPECTS OF OIL CONSERVATION
REGULATION

Chapter 1

An Approach to Conservation

Crude oil and natural gas—the principal forms in which petroleum is withdrawn from the earth—are, taken together, the raw materials for a major fraction of the energy that supports the American industrial system and for a considerable, though smaller, fraction of energy that supports other industrial economies. Although the sources of crude oil and natural gas overlap at the primary producing level, they support industrial structures of very different sorts. Natural gas, for the most part, flows directly to its end uses. Crude oil entails a highly complex structure of transport, refining, and marketing and also is the raw material of a vast petrochemical industry. Within the wide scope of the industry, the present study carves out a limited segment for examination. This segment is confined to the level of production of crude oil in the field and, at that level, to particular characteristics related to public regulation of the industry.

Because of the peculiar conditions of supply, combined with the system of property rights relating to access to underground sources, most of the states possessing large known deposits of crude oil have set up agencies that closely regulate the conditions under which oil is produced. The industry structure that results from this kind of regulation is unique among American industries, and its particular combination of private enterprise and public regulation is not found elsewhere. The structure conforms neither to the norm of competitive free enterprise nor to that of the regulated public-utility industries.

For a number of reasons, the industry has attracted an unusual

3

amount of public attention during the present century. This interest has gone through several phases. The first phase was concerned with the Standard Oil monopoly, leading to the court dissolution order of 1911. World War I made clear the extent to which petroleum was the harbinger of a new industrial revolution and how much both industrial development and national security had come to depend on this source of energy. During and after the war, the relatively small visible reserves created severe apprehension concerning the availability of future supplies and led to a strong interest both in stimulating the search for new reserves and in husbanding, under the banner of the then active conservation movement, the available supplies against wasteful use.

In the late 1920's, and particularly at the beginning of the 1930's, the discovery of prolific new sources stilled these apprehensions and led into a new phase of interest. The pace at which new supplies outran the market led both to great physical waste in the processes of production and to greatly deflated prices. At this stage the principal oil-producing states[1] initiated systems of regulatory control over production for the purpose of reducing wasteful methods of production, mediating between conflicting property rights in the sources of production, and supporting a higher and firmer price level. This was the origin of the various state regulatory agencies whose activities came to be embraced under special meanings of the term "conservation." The later structure of the industry was formed under special statutes and under the rules and regulations laid down by the agencies. It is this that has given the industry, at the crude-oil-producing level, its unique characteristics.

World War II again gave evidence of the acute dependence of the industrial system upon petroleum. For a few years thereafter, there again rose the prospect of acute shortage and special measures of public policy were invoked to stimulate the search for oil. It turned out, however, that the discovery and development of new sources during the 1950's again ran ahead of the market. Under the rules of reservoir development set by the regulatory agencies, a state of extreme overcapacity developed, and the activities of the regulatory agencies became heavily concerned with limiting the

---

[1] California, a major producing state, was not among those imposing regulations.

total amount of oil produced in order to stabilize the market and with assigning shares in the total to various producers.

In the meantime, while the regulatory agencies have struggled with the internal problems, the operations of the industry have become the object of renewed public attention in several different, but interrelated, aspects. In recent years, agencies of the federal government have been inquiring into the adequacy of the sources of energy to support the anticipated growth of the industrial structure—an inquiry that necessarily leads into petroleum as the source of a major fraction of the energy supply.

A second aspect of public interest in energy supply is the requirements for national security in various possible future states of emergency—an aspect that overlaps, but is not the same as, the requirements for industrial growth.

A third aspect of public interest arises from the fact that large reserves of low-cost oil have been discovered in other parts of the world and that oil from these sources could displace a great deal of domestic oil in the American market. This raises the question of whether or how far, in the interests of national security or for other reasons, the domestic industry should be protected against the inroads of foreign oil.

If all three of these aspects of public interest at the national level are to be discussed on a rational basis, a scrutiny is needed of the domestic industry in order to assess its potentialities for meeting future energy requirements and the nature of its contribution to national security. Since the characteristics of the domestic industry are in large degree the direct result of the rules and regulations applied by state regulatory agencies, the consequences of their activities must eventually be examined.

Policy problems of a different sort arise at the state level. These problems are essentially of two different, though interlinking, kinds. On the one hand, the economies of the principal oil-producing states are heavily dependent on the operations of the industry. States must therefore be concerned with the way in which their resources are exploited—not merely from a short-run point of view, but also with a view to the fullest contribution the industry can make to their economic well-being over time. On the other hand, a great variety of private interests, often conflicting, clamor for favorable treatment. The state regulatory systems must deal with problems of "equitable" adjustment among the various producing interests, delineating their property rights and determining the terms of their participation in the production process. These two types of problems are interlinking because state governments are under great political pressure from private producing interests, and this pressure may greatly affect the way in which they deal with the more fundamental public problem of controlling the exploitation of their resources.

Our study is concerned with the regulatory process at the state level. Its significance, however, extends beyond that level, since a clear view of the factors determining the structure of the industry is essential for rational consideration of policy problems arising at the national level.

As we noted in the preface, the study was originally prepared as the basis of discussion of the economic problems of the industry by a group whose members had many types of experience. In its present form, it deviates very little from its original form and may for that reason be somewhat lacking in descriptive background for those without such knowledge. While we were fairly clear about the limits to be placed upon this study, its title is too broad to be very informative. It does at least suggest that the focus of interest is away from technically legal, administrative, or engineering aspects of the regulatory process. Beyond that, however, we need to look briefly at the words (in their reverse order) in the title, "Economic Aspects of Oil Conservation Regulation":

Regulation. The regulation with which we are concerned is that exercised by specialized agencies set up by statutes of the oil-producing states, with some secondary reference to federal activities.

Conservation. The state statutes assigning duties to such specialized agencies are commonly referred to as "conservation laws," the agencies are often called "Conservation commissions," and the whole range of their activities in relation to the petroleum industry is commonly covered by the phrase "conservation regulation." This usage, however, diffuses the term into an almost meaningless coverage incapable of definition. Since "conservation" is a word that is used with different meaning by different people inside and outside the petroleum industry and is a word that contains multiple ambiguities, we shall have to identify the meaning relevant to our purpose as we proceed.

Oil. Oil is a primary form of petroleum, a substance that proceeds from the ground in various forms and passes through phases of transportation, processing, and marketing into a variety of ultimate product forms. State regulation is primarily concerned with the first stage of production, extended to cover the stage of entrance into the gathering network of pipelines or other transport. Our interest will be limited by this regulatory coverage. At the production level the three basic products—crude oil, natural gas, and natural gas liquids—present interrelated but somewhat separable regulatory problems. Our analysis will center on crude oil, entering into natural gas and gas liquids only where they are unavoidably related, and evading the special problems raised by federal regulation of natural gas. This emphasis on crude oil does not imply that there are no significant and pressing conservation problems relating to natural gas and natural gas liquids.

Economic Aspects. This is where the problem of definition and delimitation of subject matter becomes troublesome. Everything that a regulatory agency does has an "economic aspect." For the purpose of the present study, we shall pick and choose the economic aspects upon which to focus attention. Economic efficiency will be the central theme running through the study. We shall be persistently asking how the various elements of regulatory practice contribute—or fail to contribute—to an efficient structure of production that is in accord with acceptable economic criteria of "efficiency." In order to make such an analysis, we are necessarily committed to an extensive description of the structure of the industry and of the operating rules imposed by regulatory authority within that structure. The economic analysis will center in the costs of producing petroleum. From this central preoccupation will develop explorations in a number of different directions into the effects of regulatory practice upon (1) the incentives to exploration and development, (2) productive capacity, (3) ultimate recovery from reservoirs, (4) prices, and (5) property rights and the rules governing participation in the productive process by classes of owners.

Our analysis proceeds from what may be called a "public" point of view. That is to say, we attempt to portray the regulatory process in a way that will highlight the problems of public economic policy at both national and state levels. The purpose is not to prescribe the "correct" basis of policy; it is rather to bring into clear view the economic considerations that are relevant to determining an appropriate policy for the future.

MEANINGS OF CONSERVATION

No doubt the most commonly quoted statement about conservation was made by President Taft, who said that it is something a great many people are in favor of, no matter what it means. Since the time of the original "conservation movement" early in the century, when Theodore Roosevelt and Gifford Pinchot were the moving forces in it, the term "conservation" has been applied to many programs concerned with protecting the public interest in the development of varied types of natural resources, including forest preservation, land reclamation, multi-purpose river development, preservation of wildlife and recreational areas, minerals development, and other activities.[2] Such programs have almost always been controversial in that they touched conflicting private and public interests and involved different views on the part of national, regional, and local political and economic interests. In spite of the great diversity of situations involved, two strands of thought have been common to almost all such programs: (1) that much economic waste would occur and many social benefits would be foregone unless public regulation were introduced and (2) that long-range planning of resource development was required to eliminate the wastes and secure the benefits.

Within the broad areas to which the conservation concept was applied, mineral resources were assigned a special place because they were exhaustible and nonrenewable. The central importance of mineral resources in modern economic life, and their peculiar qualities, led to the predominance of two ideas in the discussion of minerals conservation: (1) that there should be some restriction or postponement of present use in order to have a larger supply available for the future; and (2) that economically efficient methods of recovery should be required in order to assure the fullest possible utilization of underground supplies. There is no necessary conflict between these two objectives, but it is possible in any particular circumstances to give more weight to one or to the other. The concern of both these ideas with "waste" of a resource naturally invites the attention of economists, since "inefficiency" or "misallocation" in the use of resources is one of their primary subjects. In recent

---

[2] See Samuel P. Hays, *Conservation and the Gospel of Efficiency* (Cambridge: Harvard University Press, 1959) for an interesting historical treatment of the 1890–1920 conservation movement.

years economists have built up a body of theory on conservation that centers on maximizing social benefits over time by proper distribution of resource use over time. At times economists have used the theory as an instrument for criticizing the operations of the oil industry. The use by economists of a highly technical concept of conservation and analytical procedures unfamiliar to members of the industry and regulatory agencies, has caused bewilderment, not to say resentment. It will be part of our task, as we proceed, to provide a bridge between the thought and language of the economists and the industry.

The time distribution of use of resources, regarded by some economists as an essential feature of the conservation concept, has had no explicit place in the philosophy of state regulatory agencies. Because of the multiplicity of purposes covered by regulatory practices, it is difficult to isolate a specific part of these practices that can be said to represent conservation in some precisely defined meaning of the term as it is used by the industry. The regulatory agencies and members of the industry are not accustomed to separate conservation, severely defined, from other aspects of regulation. In popular industry usage, "conservation regulation" is the behavior enforced upon the industry by regulatory statutes and agencies— or, for purposes of argument, what various segments of the industry think ought to be enforced.

In order to make a clear analysis more precise concepts are needed, and we shall attempt to supply them as we proceed. In general, our discussion of conservation will be ticd to the idea of "prevention of waste" (achieving efficiency). This phrase unfortunately contains a difficulty of its own. As Professor Mason has said:

If . . . we push this notion of waste avoidance to its ultimate conclusion the concept of conservation becomes very broad indeed. To conserve would mean to economize and the theory of conservation becomes the theory of ideal output. Or rather, since the methods of production as well as quantities of output are at issue, it becomes the theory of ideal inputs with its central concern on optimum allocation of resources among various uses.[3]

---

[3] E. S. Mason, "The Political Economy of Resource Use," in Henry Jarrett (ed.), *Perspectives on Conservation* (Baltimore: Johns Hopkins Press for Resources for the Future, Inc., 1958), p. 161.

If "conserving" and "economizing" are the same thing, no special theory of conservation is necessary. Rather, there is a need for the application of the general theory of economizing to the special circumstances and institutions that go to make up the domestic oil industry.

In view of these difficulties, the solution we adopt is not to be the slave to definitions. We shall talk about some things that "conservation" agencies do, subject them to the test of certain conventional economic criteria, and consider the reasons for departure from strict economic criteria in actual regulatory practice. Back of this "realistic" approach, however, we face a real difficulty. The economic theory, however abstract, explores the concept of "maximum social benefit." This is a concept that cannot be ignored. The very fact that regulatory bodies exist implies some effort at least to "improve," if not to "maximize," social benefits. The acts of these bodies must be scrutinized to determine what tests of "benefit" are applicable.

While we adopt a "prevention-of-waste" view of conservation, it must be noted that petroleum conservation statutes commonly set forth two goals. One is prevention of "waste"; the other is protection or adjustment of property rights.[4] We have not only to scrutinize the meaning the statute gives to the concept of waste, but also to examine how far the concern for property rights may conflict with the pursuit of economic efficiency.

END–USE AS A CONSERVATION TEST

As we noted above, one of the early meanings assigned to "conservation" in the case of exhaustible mineral resources was the restriction of present use in the interest of "future generations." Since we are largely excluding this idea, or at least considering it only in piecemeal fashion, we must take a little space to justify our procedure.

In their study, *Scarcity and Growth*, Barnett and Morse have distilled four types of waste from their review of the conservation

---

[4] Some state statutes do not specifically mention protection of property rights. However, the constitutional guaranties of "due process" and "equal protection" implicitly include this aspect of conservation.

movement.[5] These are waste resulting from: (1) not using resources in their proper order of priority, (2) failure to procure the maximum yield from nature's renewable resources, (3) failure to obtain the maximum yield of extractive product from physical resources which are destroyed, and (4) the wrong use of products yielded by natural resources. The first and third types of waste are of significance to this study since the first type relates to the general principle of cost minimization in the priority of resource exploitation, and the third type relates to mismanagement of exhaustible resources by not obtaining all that is economically recoverable. The second type of waste relates to renewable resources and does not concern minerals. The fourth type of waste—improper end-use—ties in with the future-generation argument that will be neglected in this study for the reasons given below.

In the petroleum industry and among state regulatory agencies, the end-uses to which petroleum is to be put are almost completely excluded from discussions of conservation. There is, indeed, something approaching a dogmatic taboo against mentioning the subject. This taboo embodies the limiting principle that conservation is about the production of petroleum and has no reference to its final uses. The conventional principle relating to use is that the consuming public should be entirely free to express its preferences in the market as between alternative sources of fuel and energy—that is to say, 100 per cent consumer sovereignty. We do not concur in this dogmatic reason for excluding end-use from consideration, but we do find sufficient reasons for not including end-use as a major point of our analysis.

The practice of the industry in excluding end-use from the terms of discussion highlights a special difficulty in delimiting the term "conservation." Different groups tend to establish boundaries dictated by their conception of their own interests. The dominating circumstance in the industry is the great potential excess of crude-oil supply that constantly overhangs the market. Given the excess producing capacity, the members of the industry have a compelling interest in extending their markets in any feasible direction. It is therefore "bad" to suggest either that restrictions should be placed

5 Harold J. Barnett and Chandler Morse, *Scarcity and Growth: The Economics of Natural Resource Availability* (Baltimore: Johns Hopkins Press for Resources for the Future, Inc., 1963), pp. 80–82.

on "inferior uses" to the benefit of other sources of energy or that current use should be restricted with a view to the requirements of the future. At least in the short run, the oil-producing states have common cause with the industry on this point, since their current economic well-being and fiscal supports are in substantial degree dependent upon the level of oil production.

We do not wish at this point to express any opinion on this attitude. But no single industry can enforce such limits upon the scope of discussion. There are two reasons for this. One reason is that there are other sources of energy with a contrary interest which can force the subject into the open. Thus, the members of the coal industry, damaged by the advance of petroleum uses, can challenge the principle and argue for a different one. A second reason is that public agencies responsible for policies looking toward future adequacy of energy supplies have to consider the problem.

The petroleum industry itself has not always entertained its present unanimous opposition to any consideration of end-uses. During the campaign of the Oil States Advisory Committee for output restriction in 1931–32, one of the arguments was that unrestricted production was causing an irreplaceable resource to be exhausted at too rapid a rate, was directing the use of petroleum into inferior uses, and was unduly damaging the coal industry. Thus, it was stated in the first resolution adopted by the committee on March 1, 1931, that "a fair price for crude oil is essential to its conservation so as to prevent waste of oil from diverting it to uses below its intrinsic value. . . ." On December 14, 1932, the committee presented a memorandum to President-elect Roosevelt which included the statement: ". . . petroleum at cheap prices breeds waste and competes unfairly with coal." We do not mean to imply that this was a universal industry view. It was, however, a view held by some important elements in the industry.

The immediate circumstance that called forth these statements was a great increase of production which had led to a very low price. One purpose was to curtail production as a means of raising the price. The exhaustion of supplies and damage to the coal industry from the existing production-price situation were relevant parts of the argument for curtailment. When, however, production had been restricted by state regulatory measures and prices stabilized at remunerative levels, the argument lost its relevance. Since about 1949, fairly chronic overcapacity has placed the industry against restrictions of entry into any accessible market.

The coal industry, on the other hand, has on occasion forced the subject of end-use into the realm of public discussion. In spite of their competitive positions, the two industries are not on opposite sides in all matters of public policy. Limiting the production of oil works to the benefit of coal. And the domestic oil industry can join with coal to support restriction of imported oil and higher regulated prices for natural gas. However, after suffering years of competitive intrusion of petroleum into its markets, the coal industry has in recent years been agitated by the loss of markets to natural gas, on what it considers unfair terms, for what are called "inferior uses"—especially as seasonal boiler fuel for generating steam. It has therefore supported legislation to give the Federal Power Commission control over end-use of natural gas on terms which would cut off some of the "inferior uses" competitive with coal.

While the argument over the powers of the Federal Power Commission is explicitly directed to certain immediate ends, the implications are very broad; it gives rise to the idea of a "national fuels policy" designed to introduce a "balanced fuel economy," to use phrases that have entered the discussion. A part of the argument for debarring natural gas from "inferior uses" is that limited reserves are depleted with undue rapidity—an argument that can easily be extended to petroleum generally.

On a broader front, there is a body of opinion that has strong misgivings about the rate of depletion of mineral reserves and raises the question of whether we ought not to be thinking about positive curtailment of use. In the summary of its report, *Resources for Freedom,* the President's Materials Policy Commission ("Paley Commission") broached the subject in 1952 in this way: "We must become aware that many of our production and consumption habits are extremely expensive of scarce materials and that a trivial change of taste or slight reduction in personal satisfaction can often bring about tremendous saving."[6] Professor J. K. Galbraith states the matter in these words: ". . . if conservation is an issue, then we have no honest and logical course but to measure the means for restraining use against the means for securing a sufficiency of supply and taking the appropriate action."[7]

---

[6] President's Materials Policy Commission, *Resources for Freedom, Vol. I: Foundations for Growth and Security* (Washington: U.S. Government Printing Office, 1952), p. 16.

[7] "How Much Should a Country Consume?" in Jarrett, *op. cit.,* p. 90.

The question presented is whether the underground supply of petroleum is being depleted at an "excessively" rapid rate. Stated differently and more broadly, the question is one about the "proper" distribution of the supply through time—whether some lower-priority uses in the present should be sacrificed in the interest of higher priority uses in the future.

The interest in questions of present vs. future use is dramatized by the high and mounting inroads upon the supplies of nonreplaceable minerals arising from (1) growth of population and (2) the growing per capita output of goods due to improved technology. If the exponential rate of growth in these two statistical series is projected, the possible shortfall of basic resources in the not-too-distant future looks menacing.[8] But as an argument for foregoing the use of resources now, these statistical projections may prove too much. As Professor Mason has said, "If we take such calculations seriously there is little need to talk about conservation. We might as well consume what we have now since our descendants are going to starve in any case."[9]

An opposite and optimistic reason for lack of interest in the problems of the future is the faith that the "indomitable mind of man" will always be capable of solving the problems of resource supply as they arise.

If we discard dogmatic attitudes and assume a mounting pressure upon available resources, the public interest requires some concern about the problems. However, if the matter is looked at in this way, apparently little can be said about present vs. future in the case of petroleum in isolation from other resources. Petroleum is a source of energy, and energy can only be applied to industries, or their products, that require the use of other minerals. There would be no problem of shortages of fuel for automobiles, for instance, if the number of automobiles were limited by a shortage of iron ore or of lead for storage batteries.

Moreover, petroleum is only one among a number of substitute sources of energy, such as alcohol for motor fuel, coal for steam generation and space heating, and shale oil and tar sands for all

---

[8] In highly industrialized countries there is some evidence that energy consumption will grow less rapidly than real gross national product because of the rapid growth of nonmanufacturing industries. Although this does not reverse the situation, it perhaps slows the growth rate of energy demand.

[9] In Jarrett, *op. cit.*, p. 179.

uses of crude oil. There are in addition the yet unmeasurable future supplies of atomic and solar energy.

There is, therefore, no way of making any reasonably useful assessment of the relative importance of present and future uses of petroleum that is separate from the materials with which it must be used and the substitutes for those uses. This conclusion supports, after a fashion, the position of the industry against control of end-use. But it is not based upon the industry's usual dogmatic argument for consumer sovereignty. On the contrary, the public interest would appear to require a constant concern for the adequacy of its future resource base, and this concern might call for limitations of present use. All that is argued here is that no rational decision can be made with respect to limiting the uses of energy except in relation to the whole resource situation more broadly viewed. There are at present no agencies with responsibility for this integrated analysis of the resources problem, that is, no agencies that attempt to explain and understand the interrelationships of such resources as energy, metals, nonmetallic minerals, and water. It is difficult to see how the federal government can do without such an agency if it is to deal intelligently with mounting problems of resource supply and use. We should stress, perhaps, that this is not meant as an advocacy for a national planning board for all natural resources, including energy. Free consumer choice and a free market system will presumably retain their place as the best available means for the allocation of most resources. This, however, does not diminish the importance of an agency whose function is to study and explain resource interrelationships and to identify the problems of public policy to which they give rise.

Present political attitudes lend little support to restriction of some present uses of resources in favor of the requirements of the future. One obstacle is the emphasis upon "growth" as the central tenet of economic policy, corresponding to the current aspirations for rising standards of living. At some future time readjustments will certainly be necessary both in the rate of population growth and in the per capita use of mineral resources. But there is no present disposition to face such problems of social adjustment until they are forced upon us by the pressure of events.

The other great obstacle is that a positive effort to restrict consumption in the interest of saving resources for future use would require a considerable expansion in the regulatory role of government, and the present political temper appears to be hightly anti-pathetic to any such expansion.

Perhaps a third obstacle should be mentioned. If some restrictions of present use were to be attempted, it would be necessary to prescribe precisely what resources and what uses were to be restricted. This would involve damaging curtailment of some industries and controversial value judgments over what were "inferior uses." It is not difficult to imagine the savage political in-fighting that would result.

In view of these circumstances, the concept of conservation as a policy of shifting the use of nonrenewable resources toward the future (one of conservation's valid meanings) appears to have no influential part in current discussion. The attitude of the petroleum industry on this point with respect to its own product is not exceptional, therefore, but is in harmony with the whole ethos of contemporary opinion and policy.

Actual policies cannot be entirely neutral as between present and future. Some policies, both state and federal, stimulate new capacity, tending toward the present. Other policies curtail the use of the capacity, tending toward the future. But these policies that offset each other are not in purpose concerned with matters of present vs. future. In fact, it is difficult to comprehend how the multitude of policies in the diverse natural resources areas fit with one another into any consistent whole.

Whether the present lack of concern for the future of natural resources generally represents economic sense or idiocy, from the standpoint of public interest, the present generation will probably not live to see demonstrated.

THE ECONOMIC THEORY OF CONSERVATION

In order to understand some of the later discussion that is much concerned with "efficiency," it may be helpful to summarize some lines of thought that economists apply to the concept of efficiency. Lack of familiarity with this apparatus of analysis is the cause of much misunderstanding and confusion on the part of members of the industry. Professor Stephen McDonald of the University of Texas has summarized an economic definition of conservation "as action designed to achieve and maintain the optimum time-distribution of use of natural resources."[10] He points out that Professor Mason put this in negative terms: "Conservation is the avoidance of wastes associated with a faulty time-distribution of use of natural resources."[11]

This meaning of "optimum time-distribution" is not the same as the concern for future supply discussed in the preceding section. It is, on the contrary, a way of defining an economic test of efficiency to be applied to various sets of circumstances. The central position of "time-distribution of use" cannot be understood without going back to the economist's theory of investment, or capital theory. This starts from the presumption that each individual firm is attempting to secure the most valuable stream of future income from its capital investment. Investment can be made in ways that yield future income at different time-rates. The problem is to determine which of the various income streams is the preferable one in terms of maximizing the profit. All such streams, if their size is known, can in fact be compared by the test of which one has the greatest present capital value.

Let us look at a greatly simplified illustration. Assume an oil company to own a complete, but undeveloped, reservoir of known physical properties and the owner is free to develop it as he sees fit (no regulation). This resource may be viewed as an investment; that is, it can produce revenue (satisfaction) today or at some time or series of times in the future. To get all his revenue today the owner would sell the reserves. If he maintains ownership, production can be pushed nearer the present or further into the future. Thus, the resource owner faces the problem of the *rate* at which he will produce the stock to maximize his returns over the life of the resource. There are many possible alternatives. Wells could be drilled on every quarter acre of land and each produced as rapidly as possible. This would push gross returns nearer the present, but would probably be costly to undertake. Or a single well could be drilled to drain the reservoir. This would push gross returns into the future, but would result in lower costs per unit of output. Or some intermediate timing could be followed. The resource owner's

---

10 Stephen L. McDonald, "The Economics of Conservation," paper presented to the Rocky Mountain Petroleum Economics Institute, Boulder, Colorado, June 18, 1964, p. 6.
11 In Jarrett, *op. cit.*, p. 162.

economic problem is to determine which alternatives will yield him in present-value terms the largest return during the life of the investment.

The principle at work is that, when expected future net income streams are discounted back to the present, the one adding up to the largest present value is the optimum one.[12] Such a comparison implies that future costs, prices, output, and interest rates are known. This assumption, though "unrealistic," is necessary to demonstrate the economic logic of the investment process, however imperfectly applicable it may be in practice. The principle of maximization of profits *over time* provides the economic criterion of efficiency under private competitive enterprise, subject to modification of a social character. As we shall see, such modifications are important in relation to oil. They do not impair the logic of private investment, but do add other constraints to the test of economic efficiency.

Under any program of development and production, the crucial economic calculation is to estimate the *marginal net return* (revenue less costs on the incremental unit of production) today and for each future period. By discounting the expected marginal net return for each future period back to the present, we find net returns on a comparable basis, i.e., in today's dollars. We are, in effect, comparing the current marginal rate of interest, adjusted for risk and uncertainty, to the present value of a flow of expected returns. Thus, theoretically, if the current rate of interest is less than the discounted expected return for future periods, rational economic behavior dictates slowing down the rate of production (i.e., investment) to get more future dollars. However, more production in the future results in higher expected future marginal costs and thus lower expected marginal net returns. Adjustments in the rate of production are *in theory* made until the discounted marginal net returns in each future period are equal to the current interest rate and thus equal to each other.

---

[12] There are two reasons why the concept of "net income" is not as clear-cut in relation to oil production as it ideally should be: (1) many capital costs are "expensed" in the year of investment rather than being spread over time as depreciation; and (2) much of the current revenue over and above accounting costs represents "depletion" of a capital asset: oil in the ground. However, there is a definable flow of "net revenue" which is amenable to the discounting process for purposes of comparing the present value of alternative streams of revenue.

In the words of Anthony Scott, the owner of the resource

... must now be envisaged as comparing the net return which he can obtain in each period at the margin of output from a unit of resource with the highest discounted return that an extra unit could earn in some future period.

This alternative return which is expected to be available in the future is the opportunity cost or "user cost" of production today.

The producer equates

... his marginal net revenue and his marginal user cost. When these two are equal in all periods, the owner will have succeeded in maximizing the present value of his enterprise....[13]

Using these concepts, Professor McDonald arrives at the definition of conservation quoted at the beginning of this section, "as action designed to achieve and maintain the optimum time-distribution of use of natural resources."

Stripped to its skeleton, this is the concept of conservation expressed in some recent theoretical writing. With additional simplifying assumptions, it can be applied conceptually to the aggregation of fields and companies that constitutes an industry. It can also be made to include prospective reserves as well as known reserves and thus encompass exploration activities as well as development and production. Under these circumstances, oil then becomes both a "flow" and a "stock" resource. The investment problem is really no different.

If conservation theory is looked at from the total social view rather than from the view of an individual company in the resource industry, it can be said that ideally society wants to obtain the maximum net benefits (discounted) from a resource over time. This type of calculation requires quantifying present and future expected *social* costs and *social* benefits, as well as determining an appropriate social discount rate. When social considerations are introduced into the economic concept of conservation, we encounter the concept of "external costs" which do not enter into the calculation of optimum investment and profit maximization by private firms. The preceding discussion noted that the principle of an individual oil resource owner maximizing the net income from the resource over time could be generalized to include all owners of oil. In its

---

[13] Anthony Scott, *Natural Resources: The Economics of Conservation* (Toronto: University of Toronto Press, 1955), p. 6.

generalized form we would say that the industry has achieved optimum resource use and thus that society as a whole has received
the greatest possible net benefits. There arises against this generalization the problem of "externalities." In the development and production of oil, there may be, for example, certain "costs" (or benefits) to society that are "external" to any one company in the
industry. If there are such external costs, they obviously do not
enter into a company's calculation of its own discounted maximum
net return, yet they are costs which someone must bear. For example, salt water is often produced concurrently and unavoidably
with oil. Without regulation to the contrary, the oil producer will
dispose of this water as inexpensively as possible. Perhaps he will
merely dump the water into a nearby stream, thus causing pollution of fresh surface-water supplies. Pollution is a form of social
cost that someone must bear. The oil operator, in this case, bears
little if any of this cost directly. We have regulations which in some
cases attempt to eliminate these costs or to shift them to the producer, i.e., make them internal.

Of course, the effort to quantify social costs and social benefits
under maximizing concepts must be extremely imprecise. Moreover,
social benefits are not all economic in character and many of them
are not subject to measurement on any economic yardstick. Nevertheless, in any analysis attempting to apply a test of economic
efficiency, it is necessary to work with the concepts of "cost" and
"benefit" and to ask in quantitative terms, however imprecise,
whether the costs are minimal relative to the benefits or the benefits are maximal relative to the costs.

When first encountered by men with "practical" experience in
the oil industry either as operators or regulators, the kind of economic theory outlined above sounds like pure gibberish. It appears
to make no contact with their activities, and it is expressed in a
language with which they are unfamiliar. Therein lies much of
the confusion that arises in discourse between economists and oil
operators and regulators and that creates the impression that economists are in some way "enemies" of the industry. This is illustrated in the words of one regulator: "As a lawyer and a former
member of a state conservation commission, I am convinced that
their [economists'] principle of 'economic conservation' is both
illegal under existing statutes and politically impossible in our

democratic society."[14] This, however, misinterprets what the conservation theorists are up to. They are not attempting to provide a theory which can be "applied" directly to the reconstruction of the oil industry or its regulatory processes. They are in effect raising a probing question: Does oil-conservation regulation, as currently practiced, promote "efficiency" in an economic sense? To arrive at any answer, it is necessary to have a method of defining "efficiency." Once this is done, the theorists can then (1) concede the imperfect degree to which the dictates of efficiency can be approximated and (2) admit into the realm of regulatory action other elements that may be considered socially desirable outside of, or in contravention of, considerations of economic efficiency. The function of a theory is not to tell what ought to be done, but to provide patterns of analytical thought that clarify the relationships within a complex situation.

To establish some basis for a greater degree of mutual comprehension, we must pursue this subject a little further. We will begin with an example that will not cause the practical oil man much difficulty in following the theoretical language. This is the system of oil property rights known as the "law of capture." Basically, this rule states that the owner of land who produces oil from a well on his land is recognized as having produced or "captured" it, even though a part of it is drained from the land of his neighbor. Where a number of operators are producing independently from a common source of supply without regulations of any sort, the rational behavior of each will be to drill and produce as rapidly as possible in order to avoid being drained by his neighbor. Put in terms of our simple theory of equating discounted marginal net revenues or marginal user costs over time, the marginal net revenue in all future periods may approach zero for each producer. In other words, there is no certainty that the marginal barrel (or any barrels) will be there to produce in the future because competing producers will already have captured it for their own. Under these circumstances, i.e., when future production has no assured value, the producer pushes production as much as possible to the present.

---

14 Richard C. Byrd, "Practical Aspects of Petroleum Conservation," a paper presented at the Rocky Mountain Petroleum Economics Institute, Boulder, Colorado, June 18, 1964, p. 2.

If we now add the fact that many reservoirs are production-rate sensitive, which means that the faster the oil is produced beyond some level, the smaller the total ultimate recovery of oil, the "social cost" is obvious. Indiscriminate drilling and production may result in the loss of oil reserves that could have been recovered at no more, and probably at much less, cost. Still, the high social costs in such a situation are external to any single operator in the field working under an unrestricted rule of capture. Oil conservation regulation was designed in part to prevent these external diseconomies from arising. Under the drill-and-produce-as-fast-as-you-can situation, certain *internal* costs arise also. Overdrilling is clearly a cost internal to the firm. It is evident that regulation was designed in part to reduce the costs internal to the firm as well—and unnecessary internal costs are as much a burden on society as external costs. In any effort to translate the maximization theory into social terms, these external and internal costs must be included in the reckoning.

The fact of regulation also introduces a new element into the calculation of private owners. Their economic calculations have to fall within the rules imposed on reservoir development, rates of production, and other matters. However accurate their own calculations in these terms, the regulatory agencies themselves are to a considerable extent determining the degree of economic efficiency achieved in exploiting oil resources. They may, for example, be promoting efficiency by moderating the "law of capture" or may be defeating efficiency by well-spacing rules that induce excessive investment in drilling. Criteria of economic efficiency must therefore be applied to the regulatory decisions as well as to the decisions of operators, although we must reemphasize that economic criteria may not be the only relevant ones.

Here, then, a new kind of question arises of whether the industry as organized under regulation is "efficient" in some definable sense. The question, in its simplest form, is: Are the costs incurred in producing the present amount of oil greater than they need to be? But the present amount of oil produced is itself a result of the regulatory structure. So the question has to go deeper: How would the industry have to be organized to minimize the costs of whatever amount of oil was to be produced? This leads on to the still deeper level: What amount of oil would it be "economical" to produce under any given set of technical and economic circumstances?

If it be assumed that each firm, within its limited powers of foresight, is attempting to maximize its own profit, how do we analyze the effect of regulatory practices upon the economic efficiency of the industry as a whole? In considering the three questions in the preceding paragraph in the order they are asked, we are forced to the following conclusions:

(1) If an oil reservoir is served by more wells than are necessary to drain it at any desired rate of production, it violates the criterion of economic efficiency.

(2) If a reservoir is developed by numerous surface owners when costs on units could be made lower and ultimate recovery made higher by unit operation, it violates the economic criterion.

(3) If the regulatory authority enforced rules of reservoir development that brought points (1) and (2) into conformity with the economic criterion, it would need to give little thought to the rate of production that would correspond to the economic criterion. If drilling and operating costs were minimized by the rules of reservoir development, the self-interest of operators could be left to determine the economic rates of production and the economic level of investment in development, according to their best judgment based on present and anticipated costs and prices and the applicable rate of interest for discounting future net revenues. It is unlikely that a regulatory agency would be in a better position to exercise judgment on how to maximize the present value of known or estimated reservoirs.

The actual development of reservoirs has not conformed to those principles of economic efficiency. Nor can it be argued categorically that these principles ought to be applied as a matter of policy. There may be reasons for deviating from them in accordance with other social aims or practices. The advantages of having the economic tests clearly defined is to identify the deviations from them, as a basis for considering whether the deviations are justified by special considerations. While it is impossible to imagine the actual existence of an ideal state of efficiency, in view of the uncertainties surrounding anticipations of the future, it is extremely easy to identify lines of action leading in the direction of efficiency, or the opposite, as defined by the theory. This is the precise service to which the theory can be put and to which it will be put at various points in this study.

The difficulty members of the industry have in seeing the point or meaning of the theoretical analysis lies not merely in their being unacquainted with the methods and concepts of economic theory. The greater difficulty is that many of them take for granted the institutional structure of the industry and the regulatory processes that have formed this structure. They live, act, and think within these limits. The economists are not limited in this way. On the contrary, they are probing into the structure, asking what appear to some to be damaging or impertinent questions. They ask in what respect or to what degree the industry is efficiently organized. To arrive at any answers they have to have some concept of efficiency.

Members of the industry undoubtedly have divergent views on the question of efficiency. They act in their own self-interest, for which they can hardly be criticized, and these interests can and do often conflict. Some groups in the industry, in their own interests, have urged changes in regulations that are in the direction of greater economic efficiency. Other groups have opposed such changes and have even urged change, in their own interests, that would seem to lessen over-all efficiency. Legislatures and administrative agencies respond to these diverse pressures, and the result is our present set of laws and regulations.

Most economists, we may venture to say, agree that the kind of theory we have been examining is useful as an analytical tool. But there is not necessarily agreement as to its bearing upon possible changes in the regulatory process. Some tend to support an all-out effort to place the industry on as efficient a footing as possible, making economic efficiency the dogmatic basis for economic policy. If followed, this would impose drastic changes upon the structure of the industry, as we shall show later. Others give more weight to historical and institutional factors and the pursuit of various social objectives that lie outside the efficiency concept. Among the latter is Harold J. Barnett, who is critical of the narrow view of the conservation theorists. Exploring the general problem of scarcity and its causes, he writes:

So far as I know, most professional economists who have made important contributions in the economics of natural resources tend to consider conservation as concerned with the economic problem of *time rate of use* of natural resources. . . . Part of conservation doctrine was indeed this familiar proposition. . . . But part of conservation doctrine, and the *gestalt* in which time rate of use appeared, go quite far beyond the time rate problem. . . . And it [the economic view] fails to credit conservation with an important and partly successful revolution in social ideas and applied political economy.[15]

He is saying in essence that the natural-resource industries presented real problems that were unamenable to the rules of the free market, and that there were forms of "waste," not conforming to the economists' definition, that called for new forms of social action. In this view, "conservation" is a wide enough term to cover the historical measures taken to deal with these real problems.[16] Barnett believes that the economists' critical view of the concepts of "waste" and "wise use," as found in noneconomic literature on conservation, is not always justified. He notes that these concepts of waste and wise use "are at variance with economic common sense and understanding if these are based on laissez-faire premises. But this is the essence of the matter—conservation doctrine did, in significant degree, reject laissez-faire consumer sovereignty principles.[17]

Barnett does not reject the purely economic view of conservation, but he feels it to be incomplete. He is speaking of environmental factors which the pure economic theory need not assume away or overlook, but which those using it frequently do—much to the distress of the companies and regulators who live in the environment. Barnett does not give us a nice, precise definition of his concept of conservation, perhaps largely because he feels there is no neat concept there. Conservation in his sense is more a movement and an ethic for avoiding waste in some physical sense than a mere extension of the economic theory of optimum allocation of resources between uses and through time. Many economists would agree. But there still remains the necessity for a theory of economic efficiency to bring into play when the subject under discussion precisely *is*

---

[15] Harold J. Barnett, "Population Change and Resources: Malthusianism and Conservation," in Universities—National Bureau Committee for Economic Research, *Demographic and Economic Change in Developed Countries* (Princeton: Princeton University Press, 1960), pp. 447–48.

[16] Among these problems, one might include: (1) the problem of "skimming the cream" off of our resources; (2) the puritan directive not to "waste" our God-given resources; (3) the problem of national defense and national self-sufficiency; and (4) the view, let's err on the side of "saving" our resources, since the future is really quite uncertain.

[17] Barnett, *op. cit.*, pp. 448–49.

economic efficiency. Whether or not it is called "conservation theory" is immaterial.

INDUSTRY CONCEPTS

The preciseness and narrowness of the theory outlined above is in marked contrast to the looseness and vagueness of the conservation concept to be found in industry circles. As used by the industry and its regulators, the term "conservation" cannot be defined—or perhaps one might say that there are several industry definitions. However, in order to draw some comparisons between the industry's view of conservation and that of economic theory, a brief summary is appropriate at this point, with some comments on problems about which we will have more to say later. Professor Erich W. Zimmermann, often quoted on conservation by the industry, noted that:

... there are two major objectives of the present *regulatory* program: (1) the prevention of waste of oil and gas, through which the ultimate recovery of these products from their reservoirs is greatly increased; and (2) the protection and adjustment of correlative property rights appertaining to each owner of land in an oil or gas pool. These two objectives have become the primary aims of petroleum *conservation* and *regulation*. [Emphasis in the original.][18]

Zimmermann indicates that these two objectives are "... coequals, each worthy of pursuit in its own right, one for the sake of what may be called economy, the other for the sake of equity."[19] He then notes that the industry adds two corollaries to the above objectives. Conservation as conceived by the industry applies to oil production and storage and not to its use, and prevention of waste includes "adjustment of production to consumptive needs."

Since Zimmermann never defines conservation in a more general sense, one comes away from his book only with a concept of petroleum conservation (regulation?) as it was practiced in the first half of the twentieth century in the United States. Even here he provides no test by which to judge the degree of success in meeting the two major objectives he suggests. While he adheres to prevention of

---

[18] Erich W. Zimmermann, *Conservation in the Production of Petroleum* (New Haven: Yale University Press, 1957), p. 24.
[19] *Ibid.*

waste and protection of correlative rights as the primary objectives of oil conservation, he provides no general criteria or guidelines to judge what is or is not wasteful or inequitable. While we can agree with him that need for regulation arises from the unique characteristics of oil, his reluctance to come to grips with economic relationships leaves untouched the basic question of economic efficiency.

Professor Zimmermann at least provides a lucid account of the circumstances that led to regulation and of the content of regulatory action. Nothing similar appears in any analysis from within the industry. The rather amorphous character of industry thinking on conservation is well illustrated in *A Study of Conservation of Oil and Gas in the United States, 1964,* by the Governors' Special Study Committee of the Interstate Oil Compact Commission (IOCC). Conservation as a general concept is not defined, but there are numerous statements scattered throughout the report that spell out the goals of oil-conservation regulation. The following excerpts illustrate this:

The primary purpose of a petroleum conservation statute is to prevent physical waste aboveground and underground in oil and gas production operation; however, the due process and equal protection clauses of the Federal constitution, and usually similar clauses in state constitutions, as well as provision or provisions in the conservation statutes, require that the regulation must protect the property rights of those who have the right to produce, subject to reasonable regulation to prevent waste.[20]

The implication here seems to be that waste prevention and protection of correlative rights are not "coequals," as was maintained by Zimmermann. Prevention of physical waste appears as the primary aim. At another point, the report states that "The term conservation as applied generally in the oil industry denotes practices for the greatest ultimate recovery that is economically feasible."[21] This matches up badly with the statements on physical waste and property rights. Since conservation regulations have a substantial impact on both, this statement implies that any practices that are followed under conservation regulation are conservation. Regulation itself helps to determine economic feasibility.

In another passage the report notes that, "The purpose of regu-

---

[20] IOCC Governors' Special Study Committee (Oklahoma City, 1964), p. 6.
[21] *Ibid.,* p. 47.

lation is to prevent physical waste in a reasonable, effective way, and any effect on prices and economic waste is incidental."[22] Clearly, this concept contradicts that presented immediately above dealing with the greatest ultimate recovery that is economically feasible. Conservation regulation cannot at the same time be concerned with recovery that is economically feasible and unconcerned about economic waste.

As a partial goal of regulation, the study also states:

Thus, conservation not only is concerned with good management of those resources already defined by field discovery and development, it is likewise concerned with good management of those conditions that will allow, stimulate, and further encourage the creative spirit and scientific effort of the industry toward further exploration.[23]

This is a rather interesting goal. If put in the context of preventing physical waste, it must mean that the aim of conservation regulation is to prevent the waste of undiscovered oil resulting from inadequate exploration. If, on the other hand, we interpret exploration incentives as an additional goal, this brings us face to face with the economic theory of conservation. Companies will explore if their expected future returns from exploration are sufficiently high. The conservation goal is thus, in this context, to raise expectations of future returns. This statement may be read to mean that an operator contemplating exploration can rely heavily on the "rules of the game" being maintained by regulation in the future. He can expect to share in the market and to sell at posted prices or close to them. It is questionable whether the expectation of the continuance of recent past market situations creates incentives or disincentives for exploration.

Turning from the detailed background study by the IOCC to the summary and conclusions adopted by the eight governors who signed the Governors' Committee report, we find some further elaboration of the concept of conservation. The summary states that, "Ideal conservation will result when the maximum hydrocarbon recovery is achieved within a reasonable time period and with a minimum expenditure."[24] To an economist, this definition is startlingly different from the concept of prevention of physical

---

[22] *Ibid.*, p. 92.   [23] *Ibid.*, p. 22.   [24] *Ibid.*, p. xvi.

waste and protection of correlative rights. It sounds very much like a version of the economists' theory of conservation, but in the report there is no elaboration of these ideas in an economic context.

The treatment of the conservation concept, illustrated from the IOCC report, is characteristic of almost all industry literature on the subject. Such a treatment is a standing invitation to trained economists to attempt to create some order in what appears to them to be partial statements and inconsistencies. They do this by applying criteria of economic efficiency to the operations and regulatory practices of the industry. This at once exposes the wide extent to which industry practices cannot meet the economic test and subjects such practices to an examination of the validity of the social purpose of practices which are uneconomic in character. Such examination of their familiar environment has a disturbing effect on the minds of those who live in the industry. Most of them have a stake in various regulatory practices, and they would prefer them not to be openly questioned. Therefore, they tend not to like the intrusion of economists into the arena. The economist may blithely ignore the structure and environment of the industry developed over decades, and thus be accused of ignoring "reality" in developing his theories. As we implied earlier, the economist and the industry cannot hope to solve what is a mutual problem until communications improve.

A final goal of regulation, not covered in the preceding account, is implied in much of the industry's writing on conservation. This is to maintain a strong, healthy domestic producing industry for the purposes of (1) supporting the general well-being of state and regional economies in terms of employment and income, (2) maintaining the tax revenues and thus the operations of state and local governments in producing areas, and (3) contributing to national security, or the defense posture of the entire nation. We will not explore in detail at this point how these purposes fit in with the more central goals of waste prevention and protection of property rights. They cannot, however, be omitted because they influence the thinking of regulators, in particular on such things as incentives for exploration, shutting in wells which are not needed for efficient recovery but which support local employment, and the need or lack of need for "excess" producing capacity.

In this long digression we have set a sharply defined economic theory of conservation against the strands of thought that are to be found within the industry and regulatory agencies. We shall not hereafter pay much attention to the theory as such. We shall be more concerned with concrete aspects of industry organization and regulatory practice, viewed as a process of satisfying, or adjusting, multiple and possibly conflicting goals. But the economic theory of conservation will still be visible in the background as a constant reminder that, for the purposes of assessing the consequences of regulation, criteria of economic efficiency must be available.

MANY EYES ON CONSERVATION REGULATION

The affairs of the crude-oil-producing industry are a matter of interest to a wide range of people. At one extreme is the national interest in future adequacy of energy supplies; at the other extreme is the interest of well owners in their monthly production quota. In between lie a variety of other interests, including the interests of states in their economic basis and the interests of local communities, and the academic interest of students of industrial organization. Because of the diverse facets of interest, the affairs of the petroleum industry are viewed through many sets of eyes and judged by various standards. Whatever the angle of vision, it comes eventually to focus on the regulatory authorities whose rules and regulations, as limited by statutory law and judicial decisions, determine the structure and practices of the industry. An industry of such vital importance, whose structure and operating practices do not conform to the usual American principles of industrial organization, is bound to undergo critical scrutiny. And regulatory authorities whose every act affects the economic well-being of members of the industry will always be at the center of some controversy.

At the cost of much oversimplification, the varieties of interest may be classified as those of "outsiders" and those of "insiders." The outside interest is primarily concerned with the efficient operation of the industry as a contributor to general economic well-being. The inside interest is concerned with property rights and a favorable position for participating in the opportunities for private gain.

These contrasting viewpoints do not normally arise in the examination of an American industry. It is generally assumed that in the usual case economic efficiency goes hand in hand with private economic gain, as the condition for competitive survival. This is commonly regarded as the great merit of a freely competitive system, and any marked deviation from this norm attracts attention as evidence of monopolistic features that are, or may be, contrary to public economic policy. But in the petroleum industry, survival depends upon a set of factors to a large degree divorced from the test of economic efficiency. It depends upon access to underground supplies under rules set by public regulatory agencies under controlling legislation. It depends upon market access which is significantly influenced by rules set by the same agencies. It depends upon prices maintained by restriction of domestic production imposed by the agencies and upon the restriction of oil imports by the federal government. These points will be developed in more detail as we progress.

It perhaps could be stated that the outsider calls attention to possible changes that could have the effect of achieving greater efficiency for the entire industry. The insider, normally with a substantial vested interest in the status quo, attempts to achieve maximum individual efficiency within the constraints of regulation and to change them only in ways which serve his own private interest.

The private rights philosophy is not, however, oblivious to certain claims of economic efficiency, and it is this fact that provides the common ground between the two points of view. The conventional way of describing the purpose of regulatory legislation is to say that it was designed to "prevent physical and economic waste." This is true so far as it goes. But the meaning of "waste" is seldom clearly defined. Also, the description of the purpose of regulatory action should be supplemented by an additional phrase, "with minimum disturbance to the rights of individual owners," and by an additional statement of purpose, "to stabilize prices at levels profitable to the bulk of individual operators." The purposes of regulation were never conscientiously fixed on achieving oil production at minimum cost and on promoting the maximum economic recovery of oil.

Outsiders, and particularly economists, are predisposed to see the petroleum industry as an inefficient one in which costs are made too high because of investment in unnecessary wells and over-extended producing capacity, and one in which the methods of reservoir development leave too much oil unrecovered—all of which is true by any ordinary economic criterion. Persons inside the industry are predisposed to see that the physical waste of earlier periods was overcome, the market stabilized, the expanding requirements of the market met, the property rights of individual owners minimally disturbed, equitable participation in the rewards of the industry improved, and progress made toward the more efficient organization of production—which is also true, up to a point.

Some outsiders are prone to overlook the radical changes in technological areas that make obvious to everyone within and outside the industry the "wastes" of thirty years ago. They are also apt to ignore some substantial progress in the area of regulation that has gone a long way toward eliminating wastes. On the other hand, some insiders ignore technology also, and advocate continuance of regulations that were based on the state of the arts of decades ago. They refuse to listen to any rationale of regulation that includes change. Not all outsiders and insiders, of course, hold these views.

Since proof of economic inefficiency is relatively easy to establish, the question arises of the extent to which it is possible, or desirable, to transform the industry by giving it a more efficient basis. There are serious obstacles to such a process. Whatever one's devotion to economic criteria, the structure of the crude-oil industry has been formed over the past generation by the regulatory rules applied to it. It is, so to speak, an organism that can be modified, but perhaps cannot be rapidly or drastically re-formed. At the same time, new problems are overtaking the industry that appear likely to initiate, and perhaps already have initiated, changes from within in the direction of closer conformity to economic criteria. There should, therefore, be room for useful discussion of the directions of change among those who live with the industry and those who observe it from the outside.

Chapter 2

A Historical Digression: The Road to the Interstate Compact

In the early history of the petroleum industry, individual states were beset one after another by the physical and market problems that arose under the "law of capture," and each state took experimental steps to cope with the problems through some form of regulation. By 1930–31, however, there had arisen a market situation so damaging to the interests of all states and all operators that the individual states were convinced of the necessity of taking cooperative action. Such cooperation was given added impetus by the obvious concern at the federal level and the very real possibility that federal intervention would come quickly if the states were unable to solve their own problems. No one can understand the regulatory system of today without some knowledge of the cooperative efforts of the states from 1931 to 1935. A visible outcome was the organization of the Interstate Oil Compact Commission in 1935, but it is less this outcome than the stages leading to it that need to be known.

THE BACKGROUND OF THE COMPACT

The problem of the industry in the decade prior to the Interstate Oil Compact arose primarily from three factors: (1) the sporadic discovery of great new fields, culminating in the East Texas Field in 1930; (2) unrestricted drilling and production, and (3) the depression of the 1930's. The unhappy results ran in two directions: One was the dramatically wasteful production that included surface physical waste of unmarketable production, economically wasteful excess of investment in development and production, and, most serious of all, the premature exhaustion of reservoir energies necessary for large ultimate recovery. The other direction was extreme dislocation and instability of markets and recurrent periods of very low prices. The characteristics of the industry alone would have got it into sufficient trouble. But when the flush production of the Oklahoma City and East Texas Fields coincided with the deepening depression in 1930 and 1931, the effects were catastrophic for the varied interests dependent on the industry.

Prior to this time a few states had already started to exercise some control over production and Oklahoma had pioneered in regulatory adjustments to the arrival of rich new fields, usually on a field-by-field basis, temporarily and in cooperation with the operators. After 1925, however, the "rising tide of oil" created problems that were serious and persistent and called for continuous state-wide action. The impact of new supplies may be seen from the following quotation:

In 1926 crude production in Oklahoma reached the then-high figure of about 179,195,000 barrels, an average of about 491,000 barrels per day, an amount slightly under one-fourth of the total production of the United States that same year. In 1927 Oklahoma's production increased over fifty per cent, or to 277,750,000, a daily average of about 761,000 barrels. After that year, its production constantly declined, except for one increase, until it fell below an average of half a million barrels per day, not because the pools in the state could not have produced a great deal more, for the potentials for a long time mounted higher and higher, but because other states, Texas in particular, brought in many new pools, several of which were more strategically located marketwise.[1]

As a result, the Oklahoma Corporation Commission entered a state-wide proration order in September, 1928, that was the first of its kind to be made in any state. Almost immediately, however, the great Oklahoma City Field came in, touching off a spate of "town lot" drilling, and in 1930 a more detailed system of proration was initiated that was to deal with all the problems of allocating to pools, allocating to wells, exceptions for marginal wells, equitable access to markets, and ratable take by purchasers. In 1930, Texas

---

[1] *Legal History of Conservation of Oil and Gas: A Symposium*, Section of Mineral Law, American Bar Association (Baltimore, 1938), p. 151.

also made a first attempt at statewide proration, under severe political and legal handicaps, but the Texas Railroad Commission was
almost immediately swamped by the flow of oil from the gigantic,
newly discovered East Texas Field.

During 1930–31, the system of regulation was still in fairly em-
bryonic form and was insufficiently backed by legislation, judicial
decision, and regulatory apparatus to cope with the tornado of
disturbing events that occurred in those years. In this atmosphere
of emergency, not to say demoralization, the governors of the oil-
producing states set up the Oil States Advisory Committee (or Governors' Committee) to devise a plan of cooperative action by the
states. It is through the history of this body that we can most con-
veniently trace the pathway to the Interstate Oil Compact.[2]

The committee first met on February 28 and March 1, 1931, in
the midst of the East Texas production crisis. There was no doubt
about the focus of its interest. In a rambling and repetitious resolution, containing twenty points, the emphasis throughout was on
limitation of supply to support "fair prices" that would "assure to
all producers a fair and reasonable return on their necessary investments."[3] Professor Erich Zimmermann, in his discussion of this
episode, points out that:

The first objective of this Advisory Committee was to correlate the activities of state regulatory agencies in the oil-producing areas with a view to
establishing fair state allowables and bringing prices back to a remunerative level. . . . Price was recognized as a major factor in effective regulation.
It was anticipated that prices should be such as would avoid ruinous losses
to producers and at the same time not result in any unfair prices to
consumers or in monopolists' gouging.[4]

The secondary theme of conservation came explicitly to the surface in reference to the possible loss of reserves underlying
"300,000" small wells that might have to be abandoned if market
conditions did not improve. The prospective wholesale abandon-

---

[2] The most useful secondary source for this history is Blakely M. Murphy (ed.),
Conservation of Oil and Gas: A Legal History, 1948, Section of Mineral Law,
American Bar Association (Chicago, 1949), Chap. 37. Informative documentary
material is added in *The Compact's Formative Years 1931–35*, IOCC (Oklahoma
City, 1955). The present section relies heavily on these sources.

[3] The resolution is reproduced in full in *The Compact's Formative Years,
ibid.*, pp. 3–8.

[4] *Conservation in the Production of Petroleum* (New Haven: Yale University
Press, 1957), pp. 204–5.

ment was also associated with the prospect of the survival of only a few great companies that would result in a condition of monopolistic control of the entire industry. One point only referred to technical methods of conservation: prevention of overdrilling, proper well-spacing, pressure maintenance, and unit operation.

The committee undertook (1) to recommend to states the amounts of oil to be produced in each state, (2) to advise states on proper conservation laws and administrative procedures for the proration system, (3) to intervene with importers to regulate the flow of imports, (4) to intervene with holders of oil in storage to regulate the flow to market, (5) to intervene with large purchasing companies to restore higher prices, and (6) to assume other miscellaneous responsibilities. Its final recommendation was that the the oil-producing states enter into an interstate compact in order to make permanent the program outlined in the resolution, and the committee undertook to prepare the terms of such a compact.

The heart of the immediate emergency program was limitation of production by the states. While the committee had no power to impose quotas, it recommended statewide allowables, and its influence no doubt fortified the efforts of state agencies to check production against strong pressures to the contrary. In any case, production fell during 1931 and 1932, while prices rose substantially during the last half of 1931 and until December, 1932, when the committee ceased its activities. The retirement of the committee, and the absence of its persuasive powers, seems to have contributed to the rapid market deterioration in 1933.

After meeting with the Federal Oil Conservation Board in April, 1932, the committee requested in a letter to the Secretary of the Interior the assistance of the board through continued forecasting of national and regional demand, and proposed an interstate agreement "for co-ordination of conservation measures." It also approved the Secretary's move to secure voluntary restriction of imports. In June, the committee proposed a tariff of $1 per barrel.

A legal subcommittee in May, 1931, prepared a tentative draft of a uniform conservation law to be adopted by the states. The definition of waste included "waste incident to or resulting from the production of crude oil or petroleum in excess of transportation or marketing facilities or reasonable market demand." The proposed law also gave preference in production to "wells of settled production" as against flush wells.

During 1932 much attention was given to the preparation of a proposed federal statute that would authorize an interstate compact and that provided for a joint federal-state conservation board to promote and supervise the measures desired for stabilization and conservation purpose. The prejudice against federal participation was broken down by the recognition of the advantages it might have in promoting both production control and import control.

The objectives of the proposed statute were described by the committee as follows:

Requirement of the most effective and economic use of reservoir energy

Equitable apportionment of the contents of a common source of oil or gas

Regulation and control of drilling, producing and operation methods, so as to promote maximum ultimate economic recovery and use

Retention underground of oil and gas whose production would be in excess of transportation or marketing facilities or reasonable market demand, and when required to preserve the oil pools of settled production

Prohibition of waste of all kinds, both physical and economic whether occasioned by the breach of the foregoing objectives or otherwise

Ratable taking of production of oil and gas from competing fields and from wells within the same field

Authorizing unit operation of a single oil or gas field or area[5]

This statement had none of the emphasis on "fair price" contained in the original resolution and the conservation aspect dominates the wording. But the primary purpose was unchanged. A companion bill provided for control over imports.

Supported by much of the industry and under sympathetic consideration by the federal administration and Congressional committees, the bill appeared to be on its way to passage in the fall of 1932. After the election in November, however, for reasons then unknown to the committee (and still obscure), both the American Petroleum Institute (API) and the Independent Petroleum Association of America (IPAA) withdrew their support. Since it considered the case hopeless without industry support, the committee

---

[5] *The Compact's Formative Years, op. cit.,* p. 45. Several versions of this list were proposed with slightly different wording.

ceased its activities. It may be surmised that important elements in the industry either were antagonized by the prospect of tighter political control or were convinced that the efforts to control the disordered market by state cooperation would be insufficiently effective. In any case, the efforts were suspended for two years until the end of 1934.

As its last official act, the committee distributed copies of a "Uniform Act for Oil and Gas Conservation and Interstate Compact," designed for state adoption as a correlative of the proposed federal law. The draft law contained a provision relating the curtailment of production to when prices fell below the average cost of production. The cost formula was geared to the presumed necessities of "wells of settled production" to prevent premature abandonment.

The failure to establish an effective interstate system of control left the industry open to being swept into the orbit of the industry code system established under the National Industrial Recovery Act in the summer of 1933. The "Code of Fair Competition" for the petroleum industry set up a complicated system of controls over every phase of the industry from production of crude oil to marketing of refined products. It left the internal processes of proration in each state to state regulatory authorities, but state production quotas and minimum prices were fixed by the code authority with federal sanctions. Administrative responsibility was in the hands of personnel recruited from the industry, and state governments were nowhere represented in the lines of policy and command.

The code in effect transferred the power to exercise control over the market from the hands of state authorities to those of industry representatives under federal supervision. This represented a major change, not only of practice, but of the basic philosophy of state authorities who had struggled with the problem. To them conservation was inherently a public function in the interest of "all the people." Since market controls were inextricably entangled with conservation, it appeared highly improper to hand them over to the industry group. This was particularly true since some segments of the industry, in particular the major companies, had at times appeared as "the enemy" thwarting the states' efforts.

There was not at the time much overt expression of opposition to the code from state authorities—probably because their efforts to devise an effective system of control had failed and something was better than nothing. The deep cleavage in outlook may, however, be seen in a memorandum addressed to the administrator of the National Recovery Administration (NRA) by W. M. Downing, who had been an active participant in the affairs of the Oil States Advisory Committee. Mr. Downing wrote:

No committee of the Petroleum Industry should have the slightest power to declare or enforce conservation policies which are essentially governmental, nor to act as advisors to the President or the Administrator, in relation thereto.
Conservation requires among other things:

(a) Limitation of production and imports to market demand;
(b) A minimum price for crude oil that will enable the stripper or pumping well to operate without loss, and thus preserve from economic destruction several billion barrels of oil;
(c) A somewhat higher price that will necessitate [sic] the utilization of petroleum products from their higher uses and thus avoid competition that is crippling the coal industry;
(d) Regulation to conserve the reservoir energy of oil pools, so that the greatest possible number of barrels will be produced;
(e) Adequate encouragement to the wildcatter (the important man of the industry) to find the oil pools necessary to supply the demands of commerce for years to come; and
(f) The building of underground reserves, not of five or ten billion barrels, but of fifty billion barrels or more.

The proposed Chicago code covers matters not only relating to fair competition, but includes as well, a program of conservation of the petroleum national resources for the nation and some fifteen oil-producing states.
While there is much in common between an industry code and a conservation policy, and while the same provisions may be properly included in each, nevertheless they are essentially different in purpose, and require radically different treatment and administration. There is grave danger that in the guise of a code, the oil industry or some of its conspicuous leaders will, in fact, come to determine governmental policies.

. . . .

Plain common sense requires in this that everything pertaining to conservation be eliminated from the Code, or that the President set up a committee of the highest standing (no member of which is connected with any major oil company), to advise him concerning oil questions, and to whom shall be submitted any recommendations of the committee of the industry relating in any way to questions of conservation, before action by the President or the Administrator.
The second alternative is undoubtedly preferable . . .

. . . .

The question of conservation must be speedily solved. The states acting separately have tried and failed.
The time has come for the Federal Government to act. Conservation of

oil is of paramount importance. The problem is in fact national. Nothing that this administration could do would give greater or more enduring benefit to the people than to solve the problem. The President is particularly fitted to bring about the solution and his personal leadership is essential. The government can commence to act under this oil code, provided only that its action is free from any suspicion of any influence by the major oil companies. Further action will depend upon the adoption by the President of a program or a policy for the future. The Advisory Committee as above suggested could work out the detail of such a program and present the same to the President. In fact, such committee should be constituted for that purpose.[6]

This memorandum is of interest in several respects. In the first place, it accepts as inevitable extensive federal participation in the regulation of the industry. In the second place, it reiterates a traditional American theory of public responsibility. Third, it expresses the prevailing distrust by state authorities of major companies and challenges their elevation to a role of power in policy making. And finally, it exhibits the then prevailing (and still prevailing) confusion of the interlinked objectives of conservation, market stabilization, and protection of private rights and interests.

By the summer of 1934, the frenetic activity under the Petroleum Code showed signs of imminent breakdown in the attempted regulation of market behavior under code rules. With such a void in prospect, a new burst of activity began that was concerned with providing legislation for the future regulation of the industry on a permanent basis. It was expected that any new plan would continue federal control, and various bills were drawn up to that end. Important segments of the industry favored federal legislation, presumably with the expectation that it would continue something of the code system with industry participation in the regulation of the industry. When it began to appear that any new federal legislation likely to be passed would be highly distasteful to the industry, industry support faded in the fall of 1934.

In this mixed and uncrystallized situation, two moves were made late in 1934 that had their origins in the earlier activities of the Oil States Advisory Committee. The proposed federal statute that the committee had drawn up in 1932 (see p. 37 above) was submitted, with minor revisions, for Congressional consideration. This bill authorized the formation of an interstate compact and provided

---

[6] Quoted from *The Compact's Formative Years, op. cit.,* pp. 42–44.

for a joint federal-state Petroleum Administrative Board (PAD) that was authorized (1) to determine the demand for oil in the current period; (2) to determine the just proportion to be supplied by domestic production, by withdrawals from storage and by imports; (3) to recommend to each state a quota for its production; and (4) to recommend a quota for imports. This bill was eventually sidetracked.

The second event was the calling of a meeting of governors of oil-producing states early in December, 1934, by Governor Marland of Oklahoma to reactivate the idea of an interstate compact.

THE FRAMING OF THE COMPACT

The meeting called by Governor Marland led by rapid stages to the framing of a compact. The first meeting was attended by the governors of Texas, Kansas, and Oklahoma, and representatives of a few other states. A second meeting a month later included representatives of fifteen states; and nine states were represented at the meeting of February 15, 1935, when the draft of a compact was agreed upon. Governor Marland presented a proposed compact allied in principle to the bill described. It provided for an agency representing the member states which, among other duties, would make periodic findings of market demand and recommend to each compacting state the relative part of the total demand for domestic production that could be produced without waste in such state. Cooperation with federal agencies was provided for and Congress was requested to limit imports according to a percentage relation to total demand.

A complex argument wound its way through the three meetings. This arose partly out of differences over the nature of federal participation, if any. The main source of contention, however, centered on the principle of adjusting supply to market demand with a view to sustaining prices. The matter came down to a direct contest of strength between Governor Marland, who seems to have had the support of most participants, and Governor Allred of Texas. Marland insisted upon the necessity of sustaining prices in an effective conservation program. To Allred, all talk of conservation regulation for the purpose of price control was anathema; he would consider only the matter of preventing physical waste and insisted upon a specific disclaimer of any purpose of the compacting states "to authorize the states joining herein to limit the production of oil or gas for the purpose of stabilizing or fixing the price thereof . . . ," to quote from the compact adopted.

In addition, Governor Allred was adamantly opposed to any agency or mechanism that would establish state quotas or shares of the national domestic market. In 1927 Oklahoma had had about 28 per cent of total U.S. production and was soon to experience a relative and absolute decline in output (about 18 per cent of U.S. output in 1935). Texas in 1927 had about 22 per cent of U.S. production and by 1935 had raised its share to about 40 per cent. We can conjecture that the Texas governor was fearful that state quotas would be set on some "historical" or equally unpalatable basis, thus giving Texas a smaller share than it might achieve without market sharing. It is somewhat ironic that, in the market-sharing argument that took place among states in the compact during 1960, representatives from the independents in Texas were the strongest advocates of establishment of "fair shares" among states.[7]

In the end, in order to secure any compact at all, it was necessary to bow to Governor Allred's wishes. To quote Mr. Murphy, the historian of this episode, "Marland had the votes of the delegates, Allred the ultimate power." No compact was worth considering without the participation of Texas. The compact now in force is in substance the draft presented by Governor Allred with an added passage salvaged from Governor Marland's draft that the commission may "recommend the coordination of the exercise of the police powers of the several states within their several jurisdictions to promote the maximum ultimate recovery from the petroleum reserves of said states, and . . . recommend measures for the maximum ultimate recovery of oil and gas." In the context of the debate and compromise, many persons appeared to think that this provision left a side door unlocked for the entrance of collective action on production control and price stabilization. In practice, however, it has not turned out that way.

The compact, as adopted and administered, thus sidestepped the problem with which the oil states and the industry had been centrally concerned during the preceding 5 years. This was because of (1) the resistance to extensive federal participation in the control

---

[7] Texas did not advocate a central agency allocating markets.

of the industry, (2) the unwillingness of Texas to relinquish any of its sovereign powers to an interstate agency, and (3) the general inability of the states to agree on market sharing. Since the compact omitted any mechanism for determining the shares of states in an agreed total production, it kept alive the most serious source of friction among the states. On August 27, 1935, the Congress approved the terms of the "Interstate Compact to Conserve Oil and Gas," which had been agreed upon at the meeting of February 15. The organization of the Interstate Oil Compact Commission was effected at a meeting of state representatives at Oklahoma City on September 12, 1935. The terms of the compact represented a compromise arising out of highly conflicting views on the scope and role of cooperative action among the states, in particular with respect to control of production for the purpose of influencing prices.

While the compact set the limits of activities, it did not still the conflict. As late as 1954, an account of the background of the compact prepared within the commission itself[8] stated near the beginning that "The Committee (The Oil States Advisory Committee) had prepared a plan, and a Federal Statute, providing for an Interstate Compact, far more effective than the present Compact Law," and the account stated near the end that "The experience of the present Compact Commission proves quite conclusively the necessity of a Compact Commission with broader powers, and along the lines proposed by the Oil States Advisory Committee." We have no reason to believe that this view is still widely entertained in states that are members of the compact, but the fact that the statement could be made nearly 20 years after the compact went into operation suggests the long-lingering survival of opposing views for which the terms of the compact were a compromise. In any case, the terms of the compact have remained unchanged. They limit it to the functions of collecting and disseminating information to the member states, of making recommendations to a restricted degree, and of serving as a forum for the discussion of common problems. It provides for no coercive power over the actions of individual states.

The substantively important provisions of the compact are as follows:

---

8 *The Compact's Formative Years, op. cit.,* pp. 2, 47.

ARTICLE II. The purpose of this Compact is to conserve oil and gas by the prevention of physical waste thereof from any cause.

ARTICLE III. Each state bound hereby agrees that within a reasonable time it will enact laws, or if the laws have been enacted, then it agrees to continue the same in force, to accomplish within reasonable limits the prevention of:

(a) The operation of any oil well with an inefficient gas-oil ratio.

(b) The drowning with water of any stratum capable of producing oil or gas or both oil and gas in paying quantities.

(c) The avoidable escape into the open air or the wasteful burning of gas from a natural gas well.

(d) The creation of unnecessary fire hazards.

(e) The drilling, equipping, locating, spacing or operating of a well or wells so as to bring about physical waste of oil or gas or loss in the ultimate recovery thereof.

(f) The inefficient, excessive or improper use of the reservoir energy in producing any well.

The enumeration of the foregoing subjects shall not limit the scope of the authority of any state.

.   .   .   .

ARTICLE V. It is not the purpose of this Compact to authorize the states joining herein to limit the production of oil or gas for the purpose of stabilizing or fixing the price thereof, or create or perpetuate monopoly, or to promote regimentation, but is limited to the purpose of conserving oil and gas and preventing the avoidable waste thereof within reasonable limitations.

ARTICLE VI. Each State joining herein shall appoint one representative to a commission hereby constituted and designated as the Interstate Oil Compact Commission, the duty of which Commission shall be to make inquiry and ascertain from time to time such methods, practices, circumstances and conditions as may be disclosed for bringing about conservation and the prevention of physical waste of oil and gas, and at such intervals as said Commission deems beneficial it shall report its findings and recommendations to the several states for adoption or rejection.

The Commission shall have power to recommend the coordination of the exercise of the police power of the several states within their several jurisdictions to promote the maximum ultimate recovery from the petroleum reserves of said states, and to recommend measures for the maximum ultimate recovery of oil and gas.

As will be seen, in its explicit terms the purpose of the compact was limited to the prevention of physical waste, and states were committed to an enumerated list of measures to that end. Action under the compact to limit production for the purpose of stabilizing or fixing prices is specifically disavowed. Recommendations by the commission of measures that will prevent physical waste and contribute to ultimate recovery are authorized. No limitation is put upon the power of the individual states to adopt whatever measures they wish.

Omitted from any mention is the participation of the federal government, which had been actively present in prior discussions to plan for two possible lines of action involving federal participation: (1) the determination of maximum production quotas assignable to each state, and (2) the limitation of imports. Put aside was the whole subject of market stabilization, which had appeared to many to be the crux of the problem with which they were dealing. Even further out of sight were proposals, following the demise of the NRA code, for some substitute measure that would give extensive federal control over the operation of the industry.

At the time the compact came into being, it was apparently thought of generally as a stopgap to provide a preliminary basis for interstate cooperation until the time when conflicting interests and viewpoints could be sufficiently compromised to permit the critical issues—market stabilization and federal participation—to be effectively dealt with. The fact that it has remained unchanged and that earlier proposals for more extensive and more authoritative cooperation have largely dropped out of sight marks it as something of a watershed in the history of the regulation of the industry.

By the delimitation of its function, the compact commission has been insulated from what Mr. Murphy calls "the welter of conflicting theories assigned to the worship of conservation (by groups and persons struggling for a livelihood in the petroleum industry)." It has had to justify its existence by the effectiveness of its espousal of the tenets of conservation uniquely and rather narrowly defined as, to quote Mr. Murphy again, "the preservation to the finish of the motivating drives that wring oil and gas from reluctant sands."

Within its special sphere of action the compact has operated primarily as an educative influence. It is engaged in fact-finding and recommendations concerning ways and means by which states can effectively administer intelligently conceived programs designed to prevent physical waste and to promote greater ultimate recovery. To this end, its activities have been carried on mainly through such highly competent technical committees as Legal, Regulatory Practices, Engineering, Secondary Recovery and Pressure Maintenance—to name a few of the most important. The reports and special studies by these committees have been valuable additions to the literature and have no doubt been influential in the improvement of conservation practices in the states. The compact commission could not intervene within states nor enforce any behavior upon them, but its committees could say a great deal about good reservoir engineering, about legislation conducive to better drilling, completion, and production practices, about methods of administering regulatory duties, and about a variety of other matters of mutual interest to all oil-producing states. Therein has lain its main contribution to the conservation movement.

The semi-annual meetings of the commission, composed of governors and largely attended by regulatory and industry personnel, together with the meetings of its executive committee, provide a forum for the discussion of mutual problems. This close personal association has no doubt done much to keep the states in step with one another in the loosely knit system of production control and market stabilization administered by the several states.

PROBLEMS OUTSIDE THE COMPACT ROLE

The Interstate Compact, as we have seen, did nothing directly to alleviate the problems of market instability that had been the central concern of the early 1930's. These problems continued to plague the industry and state regulatory bodies down to World War II, when the whole situation reversed itself. The problems, however, had a less life-or-death-emergency character than had been true earlier. For one thing, there was no new East Texas to foul up the situation, although there were substantial reserves discovered in the late 1930's. Most troublesome perhaps was the new production from the Illinois Basin which affected market outlets for Texas and Louisiana crude in the late 1930's and early 1940's. For another thing, most states introduced relatively effective proration systems that cut off what were viewed as excessive flows of oil into their markets. Finally, the economic situation, while still depressed, tended to improve and was no longer in the state of chaotic disorganization that had prevailed earlier. In these circumstances, the severity of the market problems to be faced differed from state to state, according to whether or not they were in a stage of rapid discovery and development and how well new oil was located in relation to markets.

Together with their problems of production control, the states inherited several other sets of problems with which they have been struggling ever since: (1) They faced the detailed problems of operating a proration system that required the statewide production to be allocated among reservoirs and wells and entailed the regulation of purchasers and transportation agencies to give effect to such allocations; (2) they had to establish the rules of reservoir development that involved considerations of efficient recovery or "conservation" and the adjustment of complex conflicts of property interest; (3) they had to regulate in detail the drilling, completion, and operation of individual wells in the interests of safety, efficiency, and the avoidance of public nuisances (the second and third problems were aggravated by rapidly changing technology); and (4) those making the laws and regulations had to work in an impassioned political atmosphere arising out of the conflicting interests, convictions, and prejudices of all those who desired to make money out of the "oil business" and all those who disagreed professionally on the best technology of oil recovery. The regulators' task was further complicated by constant appeals to the courts and the shifting grounds of judicial decisions.

Chapter 3

An Outline of Regulatory Problems

Something of the broad scheme of relations involved in state regulation of crude-oil production can be shown by a semi-historical, semi-analytical display of the factors involved. Each of these points will be analyzed in greater detail in later sections of the study.

1. Under the "law of capture" without regulation, the result was dense offset drilling, in an ever-widening pattern, in order to prevent the migration of oil from under each contiguous property. For the same reason, the incentives ran toward high flush production. This state of affairs produced surface waste from improper storage and reduced ultimate recovery by wasting or misusing the reservoir energies. It also created multiple nuisances in the way of water pollution, fire hazards, and the like.

2. The unregulated flow of oil was immediately damaging to the economic interest of owners and operators because of depressed and unstable prices, sporadically intensified by the discovery of rich fields. The fiscal interests of state governments were also adversely affected, since depressed prices and wasteful depletion of reserves damaged their revenue prospects in both the short and long run.

3. The private and public interests that were affected forced upon states the duties of regulating both the methods of production and the rate of production. Regulating the methods of production presented only engineering problems that could be dealt with for each reservoir and well separately. But regulating the rate of production created a whole witch's brew of problems. If from the outset the principle of unit operation of whole reservoirs could have been put into effect, many of the ensuing problems could have

been avoided or greatly eased. But certain historical and institutional factors stood in the way of this solution. These included the legal rights of private property and a whole tradition of individual pursuit of opportunities for economic gain. Technical knowledge of, and tools to measure, how fluids behave in a reservoir and, therefore, of how to produce oil efficiently, were also inadequate at first.

4. At the grassroots level, the problems grew out of the original legal situation in which each property owner had the right to drill wells and to appropriate whatever he could produce. This gave rise to three distinct problems: (1) limitations upon the right to drill, (2) limitations on the location of a well, and (3) limitations upon the right to produce from wells. Each of these problems has had a long history of judicial interpretation and legislative action defining the rights of owners in underlying petroleum deposits, and of regulatory action within the limits of these redefinitions.

5. With respect to production from existing wells, one problem was that of regulating in such a way as to assure the proper use of the natural driving forces of a reservoir. Another problem was that of devising some valid formula, preferably embodying an equitable principle, for sharing the contents of a pool among those who had the right to produce. The emerging principle, briefly stated, has been that of the right of each owner to an opportunity to produce or receive a share in the proportion that the amount of recoverable oil or gas in his tract bears to the recoverable oil or gas in the pool. The practical problem has been that of finding a means by which to effectuate this principle—a problem far from final solution today. Progress toward this solution has been much retarded by legislative, judicial, and regulatory actions directed toward protecting other interests that are in conflict with it. But a part of the problem is that the physical characteristics of reservoirs do not lend themselves easily to any such solution.

6. With respect to the right to drill, some limitations of this right were clearly necessary for at least two reasons: (1) to protect the correlative rights of other tract owners in a common source, and (2) to permit efficient use of the energy of an entire reservoir in the interests of ultimate recovery. A third reason, much neglected, was to avoid excessive capital expenditure upon wells unnecessary for efficient drainage of a reservoir. As late as the early 1940's, many in the industry still held the opinion that more wells got out more

oil. In addition, and for quite understandable reasons, strong political pressures reflecting conflicting interests have made it difficult to arrive at any consensus in principle as to how the problem should be resolved, and there is still great diversity of practice among the various states. Small-tract owners have had a strong adverse interest to limitation upon their right to drain adjoining tracts and thus upon their ability to drill, and their interest has been an obstacle to implementing a principle of "fair shares." In addition, local communities, whose income and employment bases are partially dependent on the number of wells drilled and the servicing and maintaining of these wells during the life of a field, have been a serious obstacle to attempts to limit drilling and to shutting in unnecessary wells.

In spite of the obstacles, the trend of legislative and regulatory practice has been toward the establishing of "drilling units" of some specified minimum size (now commonly 40 acres) which tract owners or lease holders in the unit must join if they are to share in the production from a well drilled on the unit. This avoids much unnecessary drilling expense and overcomes some of the difficulties of establishing a "fair" sharing of the product. In a degree, also, it no doubt contributes to the more efficient use of reservoir energy for ultimate recovery. The problem broadens out into the consideration of unit operation of whole reservoirs as a means to minimizing cost and maximizing ultimate economic recovery.

7. Turning to statewide problems, if production is to be curtailed, it is necessary to determine (1) the total amount to be produced in the state, (2) the part of this total to be assigned to each reservoir, and (3) the amount to be assigned within the reservoir to each producing well.[1] This is the system of control known as "proration," which has become the center of the regulatory process in most of the important oil-producing states. The problem can be approached from either end—from the top down or from the bottom up. If the statewide total is not at issue, the problem is primarily one of production efficiency—that of determining for each reservoir a rate of withdrawal and other operating conditions that will not unduly reduce ultimate recovery, and of dividing the

---

[1] Technically, the allowable is to a person, with the amount he is authorized to produce from a tract, well, or wells stated in the order given. It is common practice, however, to refer to "allocations as to wells."

amount according to some appropriate formula among the owners of producing wells in the reservoir. The statewide total would then be merely the summation of its parts.

8. In the historical circumstances of controlled production, the statewide total has, however, been an issue in most states for a number of reasons. A compelling early reason for restricted production was the low price of oil and the desire to raise and stabilize it. This was connected with considerations of physical waste, but one of the strongest motives was no doubt an improvement in the economic well-being of oil producers in general. In this situation, conflicts of private interest naturally arose, since each reservoir, and indeed each well, had an interest in keeping its share of the total as high as possible. Much of the complexity of the proration process in detail arose from the necessity of devising rules and formulas to provide some working compromise among the conflicting interests.

9. The restriction of production from existing wells proceeded side by side with policies that had the effect of stimulating both the search for new fields and the further development of productive capacity in old ones. Except when matched by equivalent expansion of the market or demise of old wells, to afford production quotas to additional wells required a further restriction upon the production of old wells. There were various reasons for this apparent paradox of policy into which we need not enter at this point. It is enough to say that the two sides of policy, working hand in hand, have created a condition of peacetime chronic overcapacity, and this condition has provided the continuing motive for restricting production in the interest of market stabilization. Adjustment of production to "reasonable market demand" thus became the controlling principle in most important producing states. Complicating this situation was a national security argument that "some reserve producing capacity" was necessary to maintain the defense posture of the nation.

10. All producing states, taken together, have a common interest in restricting national or regional production to a level that will support remunerative and relatively stable prices. This is not to say that some states are not producing at efficient capacity already. Indeed, the excess capacity has been located primarily in Texas and Louisiana. Still, all states have an interest in restrictions to market demand. But on other grounds their interests are not common since,

insofar as they have spare capacity, each would prefer to have a larger share of the total market. One state or another appears almost all the time to be jockeying to increase its share of the market; it could hardly be otherwise in a dynamic industry where new sources are developed at different rates in different states. There has been sufficient mutual restraint to maintain the system as a relatively effective market stabilizing device. Yet there is competition among states to attract operators. There is continued complaint in some quarters about state shares in the total market, but no acceptable basis has been found for determination of state quotas.

11. With the establishment of relatively effective state conservation programs, the feast-and-famine aspect of oil production was eliminated. The orderly market with higher price levels provided a type of protective umbrella for small producers and stopped what might have been a natural evolution of a more highly concentrated structure of the industry. Thus we find many thousands of companies producing with some assurance that statewide allocation systems will provide each a share of the market. Such a system is a compromise of conflicting interests and is fairly acceptable to all interests, since the small producer is given considerable protection and the large company is able to reduce and spread its risks, sharing them with the small producer in highly speculative exploratory ventures to find the oil that the large integrated company must have. However, the nonintegrated "independents" and the integrated "majors" do not have compatible interests in all respects, and they often have different views about specific conservation regulations.

12. The power of individual states to increase their levels of actual production is considerably limited by the manner in which large purchasing companies secure their supplies of crude oil in various states. The allocation of a production quota to each well in a state does not automatically provide a market for each quota. That requires a purchaser. One of the most exacting tasks of the regulatory agency is therefore that of attempting to match up purchasers and suppliers. The pipelines are for the most part affiliated with the major integrated companies, which in turn produce a substantial fraction of the crude oil. If left to themselves, the large companies would naturally allocate their purchases in two ways: (1) to the sources that were most convenient and least expensive to collect and (2) to the sources which were operated by their own

companies. The indirect and limited regulatory control over purchasers therefore takes the form of attempting, not entirely suc-
cessfully, to match their purchases with the quotas assigned to pools
and wells. The lack of complete success in the matching process is
evidenced by the occasional presence of "distress" oil in some fields
which probably would be purchased if located elsewhere.

The freedom of purchasers to buy where they wish places a
severe restraint upon the statewide total which the regulatory
agency can allocate, since, if it allocates more than the purchasers
wish to take, it has no power to compel them to take the larger
amount, but only the power to compel ratable taking among wells
within a pool. The purchasers may then exercise some choices
among their available sources of supply—a process known as "purchaser proration"—leaving some suppliers without a market for
part or all of their production quota. Such purchaser proration is
used infrequently. More often, when allowables exceed the needs
of purchasers, the purchasers appear either to buy the crude and
move it to storage or to buy the crude and work out some exchange
arrangement with purchasers elsewhere. However, under conditions
of prolonged and significant excesses of allowables over needs, we
do find purchaser proration. The regulatory agencies are aware of
the conditions that bring about purchaser proration, and thus the
total allocated production is limited by the "nominations" of purchasers who state in advance their expected purchases during the
next allocation period. The purchasers also "post" the price at
which they will buy, and this price is not subject to public regulation.

Since regulatory agencies attempt to prevent "excess" supplies,
the statewide production total appears to be determined mainly by
the aggregated decisions of purchasers, and very little by any independent decisions of the regulatory authorities. Since great store is
set upon price stability, no effort is made to increase the amount of
purchases by increased allocations which would initiate price competition.

13. The final elements in the system of regulatory controls are
those lying outside the sphere of state control and in the hands of
the federal government. One element is the responsibility assumed
by the federal government for policing the prohibition against
movement in interstate commerce of oil produced in violation of
the assigned quotas. This assists the state agencies in what might

otherwise be a very onerous task.[2] The other element is federal control over imports. Without restriction, low-cost foreign oil would no doubt make severe inroads in the market for domestically produced oil. In a substantial degree, therefore, the level at which the federal government sets imports regulates the size of the total market for domestically produced oil at current prices.

The federal government also has jurisdiction over public lands, including the prolific outer continental shelf areas of the Gulf of Mexico. The federal government is, in general, passive in regulatory matters on public lands, following whatever state regulations apply to adjoining private lands. As royalty owner, the federal government has an interest in production policies on public lands and, as general overseer of the public interest, it has a concern. Finally, the federal government regulates the sales of gas moving in interstate commerce. This, of course, relates directly to problems of gas conservation. Since some gas sold interstate is produced jointly with oil, the regulation of gas sales can have an impact on oil production and oil conservation regulations.

The system of controls outlined above is related to a mixture of diverse ends: prevention of waste (ill defined), definition and protection of property rights and composition of conflicts of interest, market stabilization, maintenance of the maximum degree of unencumbered individual initiative and enterprise consistent with the regulatory system, and general care for the fiscal and economic interests of the state and local communities. Regulation is a political phenomenon, not only subject to strong pressures from interested parties, but deeply affected by attitudes, values, and prejudices endemic in the American environment.

In view of the mixture of ends and of means thereto, there are no clear-cut criteria by which to evaluate the degree of success in achieving the single purpose of "conservation," which is in practice an undefined concept. The picture looks very different when viewed from different standpoints. Our standpoint is that of economic efficiency. But even when the economic factor is foremost,

---

[2] Effective June 30, 1965, the U.S. Department of the Interior virtually ceased its policing action in Texas, Louisiana, and New Mexico under the Hot Oil Act. The Act still remains and federal activity under it may be reinstated at any time.

there are different angles of vision. If looked at in contrast to the
"bad old days" of unrestricted drilling and production, the results
are striking and the verdict favorable. If looked at from the angle
of unrealized potentialities for lowering costs and increasing ultimate recovery, the reality is far from approaching the possibilities.
This dual angle of view should be stressed because the lack of
awareness of this point results in a lack of communication between
the advocates and critics of the present regulations. It is not incon-
sistent to say that the regulations have done both economic good
and harm in the past and may do the same in the future.

No one is likely to argue that the policies and practices heretofore in force in any state are well designed to achieve either the
maximum ultimate economical recovery of oil or economically efficient organization of production. Nevertheless, the status quo has
great attractions for many of the people involved. A tolerable, if
uneasy, modus vivendi has been achieved among the disparate segments of the industry. Also, they have learned to live with the
controls under which they now operate, and they know what to
expect. The regulatory bodies work within the confines of well-
defined statutes, rules, and regulations. The resistance to change
lies in the disturbance of vested interests, in the shock to conventional attitudes toward property rights, and in the difficult problems of adjustment that would be set both for regulatory bodies
and members of the industry if sights were set severely upon the
goals of lower cost and more efficient recovery.

This is not to say that changes have not occurred. Indeed, a reading of the history of oil conservation makes it evident that there
is continual change—sometimes faster, sometimes slower. But, as
we have said, changes in regulations are painful and thus have
lagged far behind the rapid changes in the economic circumstances
of energy production and consumption both here and abroad. We
are less concerned with the mistakes of regulations in the past than
we are with the foreseeable mistakes that must occur if present
regulations are applied to future oil development.

We are convinced that large segments of the industry would
profit from change in the direction of lower cost and more efficient
recovery. In addition, long-run benefits would accrue to the economies of oil-producing states. With a mounting national interest in
the adequacy of future energy supplies, the threat of federal intervention may help overcome the inertia of state regulatory policies

and practices. A more immediately compelling pressure comes from the prospect of mounting costs and declining incentives for new exploration. Looking to such possible developments, it is worthwhile to scrutinize the present regulatory situation in order to see what the problems of change are.

Focusing attention on economic efficiency, we shall be persistently asking what the unexploited possibilities are for more efficient operation of the industry. We shall not insist that the economic answer is always the correct answer. But, by applying economic criteria, we propose to show what aspects of regulation do—and what aspects do not—follow the precepts of conservation in the sense of waste prevention. If the uneconomic aspects of regulatory practice and industry organization are isolated and clearly viewed, the argument can then be pursued unambiguously as to whether or not they are justified. Hopefully, such analysis will provide clearer insights into the critical questions of what are the proper long-run goals of oil-conservation policy and how the public interest can best be served if there are conflicting goals that must be compromised.

Chapter 4

Reservoir Development and Efficient Recovery

Economic efficiency, about which this study is concerned, is in part a function of technical efficiency in the petroleum industry. Before we examine in detail the various aspects of regulation, it is necessary to outline in simple fashion the rudiments of oil reservoir mechanics so as to get some notion of what technical considerations are relevant.

Petroleum occurs in nature in a gaseous, liquid, and solid state. Variations in temperature and pressure can change petroleum from one form to another. The form it takes at atmospheric or at reservoir conditions depends on its chemical composition. Crude oil appears as a liquid under these conditions and may vary greatly in chemical composition and physical properties.

Natural gas, a gaseous form of petroleum under normal conditions, often contains "liquids" under reservoir conditions. These heavier hydrocarbon molecules can be fairly easily removed as the gas is produced, thus yielding "natural gas liquids" (NGL). Under some reservoir conditions natural gas is dissolved in the crude oil, which, when brought to the surface, releases the gas.

As a general rule, some gas is always found with oil. However, quite frequently gas is found without oil. In these instances it may be "wet," containing liquids, or it may be "dry," containing no liquids. While gas is an extremely important fuel source in and of itself, it is also of critical importance in the production of oil. Gas is frequently one of the major natural driving forces in an oil reservoir and thus provides a vehicle for oil production. Without some pressure differential between the reservoir and the surface of the earth, oil is virtually impossible to produce. Because of its compressibility and fluid characteristics, and coupled with high pressures, gas can be used as an effective driving mechanism to displace the liquids from the formation.

Petroleum is found in various types of geological traps. The presence of accumulations of petroleum can be determined only by penetrating the trap with a drill bit. Each reservoir is unique and the general technology of drilling and production must be adapted to suit the special conditions of a particular reservoir. Among the important variables in a reservoir are (1) the chemical and physical properties of the fluids as they occur in the reservoir, (2) the porosity of the producing formation (the degree to which the rock has space to hold fluids), (3) the permeability of the formation (the ability of the rock to have fluids transmitted through it), (4) the pressure in the reservoir, (5) the temperature, (6) the presence of connate water, (7) the presence or absence of a "water drive," (8) the presence or absence of a "gas cap," and (9) "degree of saturation" of the reservoir rock with oil, gas, and water. Where oil, gas, and water are present in a reservoir, gravity usually separates them, if the rock texture permits, so that water is on the bottom and gas is on top.

Every oil reservoir is under hydrostatic pressure which, as a rule, varies directly with the depth of the producing formation.[1] The production of oil is accomplished by puncturing the impermeable cap of the reservoir and creating a lower pressure in the well bore than exists in the rest of the formation. Under these conditions the oil and/or gas will "flow" to the well bore and even to the surface, if the pressure differential is sufficient. Whether a well "flows" depends, of course, on the other factors listed above, as well as pressure. Pumping can be used to create the necessary differential if production will not flow naturally. It is clear that the formation will produce as long as a pressure differential exists and there is something to be produced. However, oil, a liquid, has very little compressibility by itself, and thus it would take very little production of oil to reduce the reservoir pressure drastically. It is here that gas plays an important role. Gas, a vapor, is highly compressible and can expand as production proceeds, but pressure at the same time falls quite slowly as gas expands. A second form of "natural" pressure may come from the bottom or edge water which fills up the space vacated by produced oil and thus keeps up a

---

[1] There are numerous important exceptions to this generalization.

fairly constant pressure. Efficient oil recovery involves utilizing the natural driving forces in a reservoir in the best possible way and, if possible, supplementing these natural drives with "artificial" drives. "Pressure maintenance" is the term applied to the injection of gas, water, or some other substance to supplement and maintain the natural drives in a reservoir. Oil production thus is a displacement process with the gas or water, or both, filling that portion of the reservoir vacated by the oil.

Since oil production is a displacement process, the permeability of the formation and its continuity are critical factors in the ultimate recovery of oil in place. An ideal situation exists where there is high uniform permeability so that the displacing agent migrates uniformly through the formation. Irregularities can cause more rapid migration in some areas than others and can cause a bypassing and isolation of sizable pockets of oil. Efficient pressure maintenance and secondary recovery operations often depend on the skill of the technicians in determining all these relevant factors and adapting their techniques to them.

We find in oil-industry parlance frequent reference to three types of "drives": a dissolved-gas drive, a gas-cap drive, and a water drive. A dissolved-gas drive occurs where the major producing mechanism is the gas that is in solution with the oil. As the oil is produced, some of this gas comes out of solution and fills the pore spaces vacated by the oil. Initial production in such a situation is often primarily oil. However, as gas continues to come out of solution as the displacing agent, production of gas occurs early in the life of a well. Once gas production achieves large proportions, the natural drive deteriorates rapidly and oil production with it. This is basically an inefficient type of drive that usually yields a fairly small percentage of the oil originally in place.[2] In addition, the dissolved-gas drive is difficult to supplement by artificial means, although in some cases the primary drive can be significantly changed to a gas-cap or water drive.

A gas-cap drive occurs where there is a natural or artificial "bubble" of gas on top of the oil in place. As oil is produced the gas cap expands to displace the oil. The displacement occurs from

---

[2] A 10 to 30 per cent recovery is indicated in Stuart E. Buckley (ed.), *Petroleum Conservation*, American Institute of Mining and Metallurgical Engineers (New York, 1951).

above. Since gas is a poor displacing agent, the advantage of the gas cap stems largely from its maintenance of pressure and keeping gas in solution. In cases where gravity drainage can be used effectively, gas-cap drives can be quite efficient if pressures are maintained by injection. In other instances, gas-cap drives are little better than dissolved-gas drives.

Water drives are similar to gas-cap drives except that water is the displacing agent, and the displacement usually occurs from below the oil or from the edge of the oil reservoir. There must be a pressure difference, naturally or artificially created, so that the water invades the oil-saturated portion of the reservoir. A sustained natural water drive usually requires a vast aquifer. Water is a far more efficient sweeping agent and is more easily managed than gas. Water injection can, of course, be used to supplement a water drive if adequate water supplies are available. Water produced with the oil can be reinjected. Recovery efficiency, as a rule, is considerably higher than for the gas-cap or dissolved-gas drive. Recovery as high as 80 to 90 per cent of the oil in place has been reported.

Finally, it should be emphasized that many reservoirs have a combination of drives. A dissolved-gas drive is present to some degree in every new reservoir. The label given to a drive reflects the most important of the drives. It should also be stressed that an artificial drive may be quite different from the natural drive found initially in the reservoir.

With this brief description of oil-reservoir mechanics, we can now turn to a consideration of optimum oil-field development and efficient oil recovery and the regulations that influence these goals.

THE CASE FOR UNIT OPERATION

Article III of the Interstate Oil Compact reads, in part, as follows:

Each state bound hereby agrees that within a reasonable time it will enact laws, or if laws have been enacted, then it agrees to continue the same in force, to accomplish within reasonable limits the prevention of:

. . . .

(e) The drilling, equipping, locating, spacing or operating of a well or wells so as to bring about physical waste of oil or gas or loss in the ultimate recovery thereof.

The elastic clause in this passage is "within reasonable limits," but the passage does specify that ultimate recovery is a fundamental concern of a conservation program. All states have taken advantage of the elastic clause to compromise the end of maximum economical recovery in accommodation to other and conflicting ends. Some of the reasons for this will be explored at a later point, but it appears appropriate at this point to present the ideal of economically efficient reservoir development as a basis for later comparison with the actual.

It is universally recognized today in expert quarters that the efficient way to develop a reservoir in the usual case is to treat it as a single producing unit. The development programs of foreign fields held under concession agreements amply illustrate this principle. Yet in practice relatively few reservoirs in the United States have had the unit treatment, except as part of belated secondary recovery projects. This cleavage between principle and practice is something that requires examination.

In 1924 H. L. Doherty, himself a prominent oil man, attacked the industry for the inefficiency of its producing methods and whipped up a controversial storm by his proposal that the federal government require unit operation. The resulting debate, linked with concern after World War I over the possible early depletion of petroleum resources, was influential in President Coolidge's appointment of a Federal Oil Conservation Board. The board gave close attention to questions of efficient recovery methods and in its fourth report in 1930 said: "The unit idea in producing oil is bound to win out, because the natural unit is the oil pool."[3] In 1929 the API had passed a resolution endorsing unit operation.

The following was one of the resolutions passed by the Oil States Advisory (Governors') Committee at its first meeting in 1931:

Thirteenth: That measures be taken in all the oil producing states to prevent the over-drilling and wasteful production of oil and gas in those areas of new production which may be developed, holding the amount of drilling to such number of wells as may reasonably seem likely to produce said oil most economically and efficiently; and also to save and preserve the gas energy of said fields; and further to encourage the idea of operators who own tracts of land, too small to support wells individually, entering into partnerships creating blocks of sufficient size that they may reasonably and economically support a well or wells, thereby reducing and preventing both physical and economic wastes due to too close drill-

---

[3] Quoted from Robert E. Hardwicke, *Antitrust Laws et al. v. Unit Operation of Oil or Gas Pools*, American Institute of Mining and Metallurgical Engineers (New York, 1948). Reprinted with additional material, Dallas, 1961.

ing; and that *we further go on record as favoring wherein possible and feasible, the principle of unit operations* as providing the fairest and most economical possible way of producing oil and as not only effecting great savings in operations but for greater recoveries of oil and as well as rates of withdrawal that will prevent flooding the markets with new flush oil supplies that cannot readily be absorbed. [Emphasis supplied.]

As a matter of principle (and ignoring the defective syntax), the correctness of these views has continued to be accepted. Along the way we find Mr. Hardwicke writing in 1948: ". . . treating the pool or field as a unit is universally accepted as offering in most instances the ideal method of operation." S. A. Swensrud, an oil company official, wrote in 1947 (using Mr. Hardwicke's paraphrase), ". . . the industry must, in the public interest, accept as normal procedure the unitization of pools shortly after discovery as the effective way to increase ultimate recovery, to reduce expense of operations (a factor in conservation), and to resolve many perplexing production problems." These opinions have been backed by a succession of technical studies. Authorities could be multiplied, but we will cite only two.

From the standard work on *Petroleum Conservation,*[4] the following series of quotations supply the basic principles:

. . . efficient recovery from an oil or gas pool can best be assured when oil producers therein cooperate to adopt and carry on jointly the most effective recovery method for the entire pool. (p. 14)

In the case of gas-cap drive or water-drive . . . since oil must be moved from one portion of an active reservoir to another portion if the process is to be used effectively, it is the reservoir rather than the well that becomes the basic unit of production and the behavior of which must be controlled. (p. 143)

. . . although secondary recovery may be quite effective in obtaining additional oil from certain depleted fields, a two-stage operation comprising an inefficient primary operation followed by a secondary operation must be considered a poor substitute for a single recovery method that employs efficient procedures from the beginning. A two-stage operation is both more costly and less efficient. (p. 193)

. . . efficient operation of an entire pool cannot be obtained unless some method can be developed whereby the individual producers can jointly, or in cooperation with each other, pursue a common objective and carry out the particular single method best designed to recover economically the highest possible ultimate recovery from each individual reservoir. (p. 341)

---

[4] Buckley, *op. cit.*

Unit operation, the study concludes, would "facilitate the incorporation into future practices of those technical advances being made by industry through research and study."[5] The ideal plan, it states, would be to initiate unit operation from the start, based on a standard agreement prior to the drilling of exploratory wells on a suspected structure.

In its review in 1951 of a series of papers on well spacing, the Research and Coordinating Committee of the Interstate Oil Compact Commission presented conclusions that are worth quoting at length:

If the full aims of conservation are to be accomplished, individual property or lease boundary lines will be disregarded in choosing well locations.

Individual wells will become channels through which oil is expelled from whole reservoirs or producing segments rather than from separate properties. Some form of unitization of fields will be necessary, with pooling of all petroleum ownership, all driving energy, and all expenses of development and production.

It is probable that in many fields wide well spacing would result in slower recovery of all ultimately recoverable oil than closer spacing. Even though close spacing involves much higher costs some close well spacing pattern might result, through a saving in time and operating costs, in greater ultimate profit than wide spacing. All of the elements of comparative costs and revenues would be involved, and selection of well density programs would be a subject of joint study by engineers, geologists, and economists.

The importance of joint study and the application of sound principles of geology, engineering and economics to well spacing problems may be emphasized by two observations:

1. Fields with characteristics suggesting the need for close spacing for adequate and efficient drainage are the ones least likely from an economic viewpoint to justify high drilling and production costs.

2. Fields with characteristics indicating [that] low well density would adequately and efficiently drain the reservoirs have frequently been burdened with costs of unnecessary closely spaced wells.

As a matter of economics or of public interest, or both, it may be desirable to plan to either increase or restrict daily production from a single pool or in many pools, to conform with market demand. If, during the development stage of a field, market demands do not require a high level of production the wells may be widely spaced according to strictly geological and engineering dictates. If markets call for high field production it may

---

5 *Ibid.*, p. 294.

be that low production from many wells will accomplish the ends of physical conservation more efficiently than high production from a few wells.

Finally, dictates of economics, influenced at times by those of expediency are, and properly should be, the most important influences in fixing well spacing or density in any field. Public conservation authorities are always in a position to safeguard public interests by refusing to permit strictly economic control of development and production practices regardless of whether or not such control would or would not lead to reasonably complete production of all practicably recoverable oil.

The ends of conservation and the demands of economics would be fully served if fields or pools could be originally developed on wide spacing patterns to determine the field limits and the reservoir and fluid characteristics. Following the studies thus made possible infill wells could be located and drilled to provide adequate reservoir drainage and to meet the requirements of conservation, economics, or expediency.

While this has been primarily a study of the well spacing problem it has necessarily included discussion of orderly oil production, involving control of production rates where necessary to conserve reservoir energy. Production control has frequently been considered to involve two possible stages of oil recovery, referred to as primary and secondary. Because the two so-called stages require the application of the same primary types of driving energy, modern thought is that there is no real necessity for dividing operations into stages and delaying the initiation of one phase of operation until another is nearly completed.

A field with an active water drive, showing relatively stable reservoir pressures, normally will not require fluid injection to maintain reservoir energy at efficient producing levels.

Limited water-drive fields or combination water-drive and gas-drive fields may require reinjection of produced brines and excess produced gas early in field life if reservoir energy is to remain at efficient producing levels. Such reinjection would tend to sustain producing rates and prolong flowing life of the wells.

If the producing mechanism is solution gas, gas-cap, or gravity and it is determined early in the life of a field that artificial means of supplementing energy are desirable, pressure maintenance, rather than repressuring, is in order. Either gas or water injected into a reservoir during the early producing life as a pressure maintenance project is much more effective than injection at a later time when reservoir pressure has declined. If early pressure maintenance is practiced the overall field control should be considered as primary, and so-called secondary recovery would lose its identity.

In conclusion, it is the opinion of this committee that, disregarding the element of time, there is not necessarily a relationship between well density and ultimate recovery from a reservoir. Rather, the ultimate recovery of oil is dependent upon the early application of good conservation practices.[6]

This passage demonstrates that the problem of efficient recovery is not solely one of engineering, but also has an economic side in relation to the profitability of faster or slower recovery. But in either case the economically efficient rate is dependent upon proper engineering administration of the driving forces for the reservoir as a whole.

The generalization that unit operation is technically and economically the most efficient way to develop a reservoir should be qualified. There are some oil reservoirs in which the development and production would be little or no different with or without unit operations. For example, in some types of dissolved gas reservoirs efficient production can be achieved with spacing and production controls. There are no doubt such cases in which pressure maintenance cannot be done because of the inaccessibility of gas or water for injection. There are fields made up of numerous pools with fairly high permeability but which are separated by less permeable portions of the producing formation. Even though these pools may be pressure-connected to some degree, the interdependence may be so slight that unitizing the entire field would be made impractical. A common industry rejoinder to critics who urge complete unitization in the industry is that there are many fields that cannot be unitized. An estimation of how many fields could be unitized would not be simple to determine, since we must include fields in all the various stages of depletion and developed in all sorts of different ways. For example, a field with an original gas-cap drive may have had its gas depleted too early through improper drilling and production practices. It may be too costly to attempt to recreate the gas cap with injected gas and there may be no possibilities for a water drive. Earlier, it might have been quite feasible to maintain a gas cap, but that alternative is no longer a real one. Thus timing becomes important in any discussion of unitization.

---

[6] *Well Spacing*, published and distributed by the IOCC for the Commission's meeting at Fort Worth, Texas, September 10, 1951, pp. 56–57.

PHASES OF INTEREST IN UNIT OPERATION

The interest in unit operation has gone through a cycle. The experience of World War I raised a question of national defense and the then existing knowledge of reserves raised a specter of rapid exhaustion. In these circumstances, the grossly wasteful methods of operating wells and developing reservoirs attracted attention and the issue was dramatized by Mr. Doherty's campaign and the controversies arising therefrom. The Federal Oil Conservation Board, appointed by President Coolidge, kept the subject in the forefront of its consideration. The idea of unitization began to dominate the planning for development of oil-bearing lands in the federally owned domain. Companies considered the possibilities seriously. Harmony on the unitization issue was not by any means complete. During the middle 1920's there was a sharp debate within the API over unit operations and the possible encroachment of federal control over drilling and production.

This wave of interest began to be pushed aside, beginning about 1930, by more urgent problems. The production race unleashed by the discovery of great new fields led to drastic action by the states to limit production in the form of proration orders limiting total state production and consequently production in individual fields. Taken together with other regulations covering the drilling and operation of wells and reservoir management, these measures attacked the most obvious and blatant sources of waste. At the same time they made substantial progress toward what must be regarded as their primary purpose: market stabilization. There thus evolved an improving situation in which much was accomplished of a conservation nature along lines that did not arouse the dogs of controversy about unitization.

As Robert Hardwicke reports, "The opinion was quite general about 1940 that a splendid job of waste prevention had been done, or at least could be done, by proper regulation based on the usual type of conservation statutes, such as regulating the spacing of wells, fixing of gas-oil ratios, controlling the amount of production, preventing the loss of gas from gas wells; consequently, it was frequently said that the necessity for unit operation no longer existed."[7] Hardwicke characterizes the state of mind that removed

---

[7] *Antitrust Laws et al. v. Unit Operation of Oil and Gas Pools, op. cit.*, p. 162.

unit operation from the focus of interest when he wrote, "... the regulation which took place under conservation statutes resulted in so much improvement ... that the defects and iniquities were accepted with a remarkable show of tolerance and resignation." [8]

During World War II there arose a new interest in potential sources of energy for national defense and, since the war, the subject of future availability of energy for a growing national economy has again come under wide discussion. In this discussion the subject of unit operation has again come to the fore. There is, however, a difference between the earlier and the more recent phases of discussion. In the earlier phase, the most active discussion took place within the industry in its industry associations and legal, geological, and engineering adjuncts. In the later phase, while certain segments of the industry retain an active interest, more discussion has proceeded among public agencies and other outsiders concerned with policies to assure future availability of energy. The industry interest shows up in connection with the development of particular fields, but there appears to be little support for forced extension of the practice. At earlier periods, and to some degree ever since, important facets of the industry have argued against compulsory unitization, whether by federal or state authority, with special and vehement aversion to federal authority except for the brief NRA code interlude.

Here again, however, it is difficult to generalize "industry opinion." It would seem that in recent years the "majors" in the industry, while not actively beating the drum for unitization, have not thrown their influence against it. In some states such as Oklahoma and Louisiana there appears to have been substantial support from "independents" for the type of "compulsory" unitization statutes found in those states. Royalty interests and particularly landowners have been serious dissenters there. In Louisiana, the regulatory agency has taken the lead in urging stronger unitization statutes. In Texas, the independents have generally been opposed to "compulsory" unitization and claim that a voluntary basis gets all that is needed done. This is also reflected perhaps in the complete silence of the Texas Railroad Commission on the conservation merits of unitization and its failure to push for legislation.

---

[8] *Ibid.*, p. 107.

THE BASIS FOR COMPROMISE OF PRINCIPLE
AND PRACTICE

On various occasions, the Legal Committee of the IOCC has proposed state statutes giving some compulsory powers to regulatory agencies—but powers only compulsory against a minority when desired by the preponderant interests. Given the effective opposition to compulsory unitization, the proposals by the Legal Committee of the IOCC have represented the most advanced compromise that appears to have any chance of being acceptable to a majority in the industry or to state authorities, and even this has been adopted by only about 18 states.[9] Approving unit operation in principle, in its most recent "Form for an Oil and Gas Conservation Statute" (1959) the committee presents a plan for compromise between cooperative and compulsory procedures.

In spite of the unassailable authority for unit operation as a conservation measure, no state requires it for oil, except by the use of limited compulsory powers along lines proposed in the IOCC model statute. This statute deals with the matter (in part) in the following manner:

7.1 The Commission upon its own motion may, and upon application of any interested Person shall, hold a hearing to consider the need for the operation as a unit of one or more pools or parts thereof in a field.

7.2 The Commission shall make an order providing for the unit operation of a Pool or part thereof if it finds that:

7.2.1 such operation is reasonably necessary to increase substantially the ultimate recovery of Oil or Gas; and

7.2.2 the value of the estimated additional recovery of Oil or Gas exceeds the estimated additional cost incident to conducting such operations.

7.3 [details of making and administering the plan]

7.4 No order of the Commission providing for unit operations shall become effective unless and until the plan for unit operations prescribed by the Commission has been approved in writing by those Persons who, under the Commission's order, will be required to pay at least _____% of the costs of the unit operation, and also by the owners of at least _____% of the production or proceeds thereof that will be credited to interests which are free of cost, such as royalties, overriding royalties, and production payments, . . .

---

[9] Texas, New Mexico, Kansas, California, and Wyoming are conspicuously absent from this list.

No state has been willing to go further than this.[10] Among the seven largest producing states accounting for 85 per cent of production, only Oklahoma and Louisiana have statutes of the IOCC type—Oklahoma requiring 63 per cent agreement and Louisiana 75 per cent agreement of both operators and royalty owners. The Oklahoma statute dates from 1945 (revised in 1951), the Louisiana statute from 1960. The IOCC type of statute is limited in that the record upon which to apply it cannot be made until the pool has been developed to a considerable extent. As a consequence, most unit operations are put in for secondary recovery. A fairly small but growing number of fields are unitized for pressure maintenance.

The 1964 IOCC conservation study throws the weight of the Interstate Oil Compact behind the efforts to get a greater degree of unitization. A few passages from this study will convey the views of the IOCC Governors' Committee.

Where unit operations are not practiced and wells are drilled and operated in a way to give each owner the opportunity to recover the share of the production from the reservoir to which he is entitled, the wells rarely are operated to bring about maximum recovery from the entire pool.

Maximum conservation can be obtained if the principle of pooling for well-spacing units is extended to consolidate into one unit all the separately owned tracts within a reservoir.

Such operations covering all or a portion of a reservoir are referred to as "unitization," and many times represent the ultimate in conservation practice.

Unitization of a reservoir can be accomplished at any time. As a practical matter, mineral rights in land are often consolidated prior to exploratory drilling under a form of agreement that gives to all participant owners an undivided interest in all production that may be obtained from the unitized area.

Voluntary agreement is the most desirable way to accomplish unitization. Delays and difficulties sometimes are encountered in obtaining complete voluntary agreement. Statutes have been enacted in several states that effect unitization by order of the regulatory agency.[11]

From this discussion, several significant points are made. (1) Unit operations often represent the ultimate in conservation practice;

---

[10] The State of Washington is the sole example of a statute in which no ownership percentage is required. The regulatory authority can, by itself, force unit operations. In 1965 there was no oil production in Washington.

[11] IOCC Governors' Special Study Committee, *A Study of Conservation of Oil and Gas in the United States* (Oklahoma City, 1964), p. 61.

(2) unitization is possible for exploratory purposes and at any time thereafter; and (3) voluntary unitization is preferable to compulsory unitization, but the latter, it appears, must be available.[12] Strangely enough, while the text of the report as well as the summary and conclusions point out that unitization is desirable and that voluntary plans can be blocked by small minority interests, there is no concrete recommendation for compulsory (statutory) unit operations. The recommendations concerning unitization are:

Each state should do all within its power to encourage and make possible the conduct of unitized operations in reservoirs or portions thereof where found beneficial.

Each state that does not have an adequate unitization statute is urged to consider the need for enactment of such a statute.

All unitization statutes should specifically require that the regulatory agency protect the correlative rights of the minority owners, as well as the majority owners.[13]

As a succinct reason for the limited use of state powers for this and other purposes, the following is offered in *Petroleum Conservation*, ". . . the deficiencies and limitations of regulation all arise from the conflict between individual property rights and the physical requirements of efficient oil and gas recovery."[14] This does not, perhaps, state the matter with quite the right emphasis. The "rights" that conflict with "physical requirements" are not indefeasible rights. They are realms of freedom of action that exist simply because they have not been restricted by law in the interests of more effective conservation. Very commonly they run counter to other sorts of rights that can be most fully protected under unit operation, such as an "equitable" opportunity of owners to share in the product at minimum cost. The toleration of inefficient methods of development is therefore a political phenomenon that is based on widespread opposition to such far-reaching interference with the tradition and practice of individual initiative and is fortified by the special interests (real or imagined) of particular groups of owners who

---

[12] The IOCC study uses the term "statutory unitization" for what we have termed "compulsory unitization." Their term is more descriptive, but we shall bow to industry usage in keeping the word compulsory.

[13] IOCC Governors' Special Study Committee, *op. cit.*, Summary and Conclusions, p. xviii.

[14] Buckley, *op. cit.*, p. 271.

fear the power of administrative authorities to favor some at the expense of others.

When one considers the fact that unit operation, in the usual case, makes production more profitable, a slight mystery arises as to why it is not warmly embraced by owners as the only sensible plan of action. One part of the answer is no doubt that while it is true *in the aggregate*, it will not seem obvious to each individual owner in the light of his own particular circumstance. But the whole answer is much more complicated than that.

Obstacles to unit operation by voluntary agreement were discussed at some length in *Petroleum Conservation* in 1951[15] under the following headings:

1. Pride of ownership
2. Pride of operational control
3. Lack of reservoir data
4. Structural or other advantages
5. Profitable obstructionism
6. Prohibitions in leases
7. Widespread ownership of royalty
8. Mistrust
9. Fear of reduced production
10. Legal obstacles.

Robert E. Hardwicke, dealing with the same subject, said in 1948 that "... years of experience have shown that voluntary agreements for unit operation for such purposes are hard enough to obtain under the most favorable circumstances"[16] and gives his own somewhat overlapping list of obstacles. In his later edition of 1961, Hardwicke thinks that the obstacles have become somewhat less obstructive, but the limited application of unit operations, particularly early in the life of new fields, suggests great continuing obstacles.

In moving from a voluntary to a compulsory plan, the problem of composing all the varied interests would be shifted to the regulatory authorities, who might well shudder at the thought of undertaking the responsibility. Possibly, however, the difficulties seem greater than they need be in practice. If development or production activity were not permitted to proceed until an agreement on unit operation had been reached, no doubt a great capacity for

---

15 *Ibid.*, pp. 288–92.
16 *Antitrust Laws et al. v. Unit Operation of Oil and Gas Pools, op. cit.*, p. 164.

voluntary cooperation would be revealed.[17] Orders similar to this have been used in a few special situations.

In talking with members of the industry and with the personnel of regulatory agencies, the authors of the present study have encountered certain recurrent themes that seem to us to express the deepest sources of opposition to unit operation, especially if compulsory:

1. The independent operators like to think of the industry as an example of "free competitive enterprise" that offers opportunities of independent status as operators to people like themselves. Officials of some regulatory agencies, springing from the same social roots, commonly entertain similar opinions, and this is probably also true of members of legislatures. One high official of a regulatory commission dismissed compulsory unitization as "socialistic." In fairness, it should be noted that some regulatory officials have been strong advocates of unit operation.

2. Unit operation would place administrative responsibility for a whole reservoir in the hands of a single person or company who, it is generally anticipated, would be selected from among the major operators with large holdings. The deep distrust that the "independents" have of the good faith and fairness of the "majors" in such a situation comes clearly to the surface. The history and tradition of the industry is one of a struggle for competitive advantage by different interests, and the idea of impartial fairness in administration is met with skepticism. The feeling seems to be that adequate statutory safeguards would be difficult to write.

3. Some operating interests, independents and majors alike, appear reluctant to admit the capacity and willingness of many regulatory agencies to exercise intelligence and impartiality in formulating and supervising compulsory unitization plans. They either doubt the ability of officials to cope with the troublesome problems of "fair" apportionment, or they question the possibility of a scientifically acceptable definition of interests, or they are skeptical of the impartiality of officials who change quite frequently and are subject to severe political pressures.

4. The industry is full of customs, usages, rules, and habits of

---

[17] Although there may be substantial legal barriers to such a solution, the barriers do not seem insuperable.

thought that are accepted as normal. The industry has attained its form and structure for the most part without unit operation, so the subject is largely academic except for reservoirs yet to be discovered and developed and for some reservoirs for which secondary recovery is appealing. Within the industry, some segments display a willingness to move in this direction under the mounting pressure of high costs, but the political obstacles are still very great. Thus, viewing conservation in its economic sense, it is necessary to recognize that the states in their attitude toward unitization have followed a policy of "nonconservation" to the extent of failing to enforce the most economically efficient methods of recovery. It is encouraging to note that more states are recognizing this problem and are passing unitization statutes. Also, there has been an increase in the numbers of such projects not only for secondary recovery purposes, but also for pressure maintenance early in the life of fields. But the obstacles to unitization appear still to dominate the situation.

Some Economists' Opinions

Taking their cue from a number of respected petroleum engineers, geologists, and lawyers who have pointed out the obvious benefits from unit operations, a number of economists have recommended that unitization be made compulsory (and much more so than existing statutes that require 63 per cent to 75 per cent of ownership interests to agree) and that it be vigorously applied. The view of this group of economists is that a major barrier to cost reduction is the separate operation of each separately owned tract, and that unitization would effectively eliminate this weakness, namely, overdrilling and overcapacity. Some members of this group also advocate a withdrawal of market demand proration regulation with the idea that, with unit operations, the free market would determine the most economically efficient levels of production and prices. Some have also advocated the elimination of maximum efficient rate of production (MER) proration (see Chapter 8 for a discussion of MER) in order to allow operators of units to make production decisions based on present and expected future economic factors.

E. V. Rostow, professor of law and economics at Yale, came to the following conclusions in 1948:

The author is one of those who have come to the conclusion that the rule of capture has proved, for this perhaps among other reasons, a socially undesirable rule of law, and that it should be changed as the root idea of our system of oil law. The preferable way to change it, however, is to impose unitary operations on the fields, rather than to undertake further experiments with the cumbersome, expensive and unsatisfactory plan of prorationing. The compulsory operation of all fields as units of production could be accomplished by requiring the organization of companies or cooperatives in which all surface owners would share on an equitable basis, either in proportion to their surface ownership or to the richness of underlying deposits. Oil production under such units could altogether eliminate the possible wastes associated with offset drilling and the other consequences of the rule of capture, as well as the many geologists' criticisms of the administration of proration laws. It would for the first time permit the number of wells to be kept to a minimum, and the flow from individual wells in the field to be determined by geological criteria rather than the accidental pattern of ownership of the land over the oil. The unitary operation of the oil fields is the only course of action which as a practical matter could permit high standards of conservation practice to be seriously followed.[18]

He then proceeds to recommend, "The only form of production control compatible with the public interest would be a federal statute calling for compulsory unitization of the oil fields."[19]

Rostow's plea for a federal statute stems from his fear that state commissions are apt to take on the views of the industry being regulated. He recognizes the problem of domination of units by major companies and advocates as a remedy the divorcement of producing operations of major companies from transportation and refining operations.

Professor James E. Nelson, writing in 1958 about the relative inefficiency of the crude-producing industry and the stimulation of the industry attempted by tax and regulatory measures, states that, "If we are forced to give this group [domestic crude-producing industry] special tax treatment, then let us restrict percentage depletion to oil royalties [sic] from unitized fields only and provide that these fields must be unitized before drilling starts."[20] He does not elaborate on this point and does not discuss the problems of

---

[18] *A National Policy for the Oil Industry* (New Haven: Yale University Press, 1948), p. 45.

[19] *Ibid.*, p. 119.

[20] "Prices, Costs, and Conservation in Petroleum," *American Economic Review*, Vol. 48 (May, 1958), p. 514.

forming a unit for an undrilled field. We shall return to this issue later.

Professors DeChazeau and Kahn explore in some detail the pros and cons of compulsory unitization and make the following finding: ". . . despite probable widespread opposition and important administrative difficulties, we strongly favor mandatory unitization for all producing pools throughout this country under federal law. Production control should be limited to prohibition of output above MER and minimum well spacing should be required of every unitized operation; conceivably, these goals might well be realized through self-interest without intervention of public authority."[21] Their feeling is similar to Rostow's, although their economic analysis is less polemic and more reasoned.

We may also mention Professor Paul Davidson and Professor Stephen L. McDonald in this group. Davidson notes that "The advantage of such an operation [compulsory unitization] would be to void the rule of capture and allow the flow from the individual wells to be determined by other economic criteria."[22] In discussing the problem of waste in a historical context, McDonald notes that, "I think it is safe to say that all of the significant evils of unregulated petroleum production sprang from nonunitized operation. . . . We could have made unit operation mandatory and left conservation, like investment, in the hands of private individuals who would be free to make flexible adaptations of well density and time-distribution of output to the ever-changing circumstances of a dynamic economy."[23]

There is some recognition by the economists of the fear of domination of producing units by major operators that is perhaps paramount in the minds of nonintegrated companies. In a unit that has multi-ownership there could arise the question of the optimum production rate from the standpoint of the various interests involved. Conceivably, some of the interests would want higher rates than others because they have markets or outlets for the crude that some of the others do not have. Or some might merely prefer to

---

[21] M. G. DeChazeau and A. E. Kahn, *Integration and Competition in the Petroleum Industry* (New Haven: Yale University Press, 1959), p. 252.

[22] "Public Policy Problems of the Domestic Crude Oil Industry," *American Economic Review*, Vol. 53 (March, 1963), p. 97.

[23] "The Economics of Conservation," a paper delivered at the Rocky Mountain Petroleum Economics Institute, Boulder, Colorado, June 18, 1964, p. 16.

receive income sooner rather than later. Or a large ownership interest might wish to hold a reservoir in a standby position. These problems would not arise if each reservoir were operated as a single separate entity as to its entire production and marketing operations. Such, however, is rarely the case.

Implicit in the compulsory unitization recommendation is the desire to have oil produced from the most efficient sources—efficient, we should add, from the purchasers' viewpoint. The sharing of markets by all producing wells would no longer be a governing principle. This aspect of the proposal probably would be most unacceptable to a large number of small producers who have relatively high-cost and disadvantageously located wells with respect to markets. A corollary to this is the potential problem of major integrated producing companies favoring fields in which their own wells are located. A discussion of this point brings us to the price structure and pricing mechanism for crude oil that we shall defer until Chapter 8.

Finally, there is a special problem connected with the process of unitization early in the life of a field. The establishment of shares among interests in a pool is difficult without fairly detailed knowledge of the reservoir and the amounts of oil recoverable under each lease in the pool. Presumably, a compulsory unit could be formed with arbitrary shares set by a commission that would then review the shares periodically to determine their appropriateness and make changes if necessary. Overpayments and underpayments could be adjusted from future production with some discount rate applied to take care of interest costs on these payments. "Sticky" situations might arise from overpayments to properties that had no anticipated future recovery, but even these could be minimized by a pooled fund to allow for such problems. The point is that oil lawyers could probably devise regulations that would provide reasonable treatment of all interests in a unit even if the unit were formed immediately after discovery.

The discussion of the problems of unit operations tends to be muddled by the failure to isolate different aspects of the subject and discuss them separately. To get the various aspects in proper focus and interrelationship, two forms of isolation are required. First, discussion of the economics of unit operations in ideal terms must be separated from the practical problems of introducing unit operation into an industry already organized on present lines. Second, the problems of introducing unit operation early in the life of a reservoir must be separated from those of unitizing developed reservoirs for purposes of secondary recovery. We may clarify these points in the following ways:

1. Ideally, as we have seen, the economic case for unit operations is that it minimizes the investment cost for development of many reservoirs and in the usual case also increases the ultimate recovery. To this case, economists commonly add the argument that, if all reservoirs were unitized, probably the whole apparatus of production control and proration could be dispensed with. If unit operation had been universal from the start, the level of production could be left to the application of business principles within a system of free markets.

When "practical" oil men encounter this argument, their instinctive reaction is one of disbelief and opposition. Part of this attitude is based upon their habituation to the existing structure of the industry and the regulatory rules applied to offset overcapacity by allotting shares in the permitted production. They are inured to the idea that supply conditions in the industry are inherently disorderly and have to be regulated. They have not, like the economists, given any thought to an alternative structure of the industry where this might not be true. The idea of an unregulated market under any conditions therefore seems preposterous.

Granted that the conditioned reflexes of the "practical" oil men are no doubt largely the result of habituation, there would nevertheless be grave problems to be met if a policy were to be adopted of requiring unit operation of new fields to be developed hereafter. The central problem would be that of adjusting the productive operations of the unitized reservoirs to the facts of production control under the existing system of prorations (which is described in detail in Chapter 6). If it were desired to favor unitized projects, it would be necessary to penalize other categories of wells with lower production allowables. In principle, if a serious effort were being made to transform the structure of the industry to a more efficient basis, unitization of new reservoirs would have to be accompanied by the phasing out of high-cost "marginal" wells. Clearly, any such transition would have to be carried out under continued regulation. A serious effort in this direction would obviously raise very controversial issues rooted in conflicts of interest.

2. Unitization of already developed reservoirs is an entirely different matter. Its primary purpose is to facilitate an efficient secondary-recovery operation. Here none of the economies of initial investment are involved, since capital has already been sunk in the original pattern of development. Compulsory unitization as a precondition to initiating secondary-recovery projects would run into great difficulties in many situations and would probably debar some otherwise useful projects. At the same time, projects that would be fruitful are often prevented by dissenting property interests, and much more effective secondary recovery would probably be possible in the long run if regulatory agencies possessed, and would use, discretionary powers to require unitization in appropriate circumstances.

The dialogue on unit operation constantly runs into cross-purposes because of the failure to keep these aspects firmly separate. The economic case for unit operation of new reservoirs is unassailable, but any discussion of how the system might be gradually transformed in that direction cannot avoid consideration of the difficulties. These are not insurmountable in any technical sense, but the solutions are not simple and the opposition is fortified by a vast weight of vested interests and ingrained attitudes, and of skepticism even on the part of those who have no reason in principle or self-interest to be opposed.

In examining the subject of unit operation, we have been unable to find any factual information concerning the extent to which it is practiced, in the sense of unitization early in the life of reservoirs as distinct from secondary recovery operations. There is, it is true, a report issued by the IOCC in 1962 on *Unitized Oilfield Conservation Projects,* which records 1,550 projects and production of 400.5 million barrels in the United States in 1961. The report contains no explanatory text, but we judge that the projects listed are mainly secondary recovery projects and many of them partial units as against full reservoir projects. While no statistical information is available concerning the extent of unit operation initiated early in the life of reservoirs, it may be assumed to occupy a very small place in total production.

One major statistical gap, which, if filled, could settle many of the arguments over the benefits or lack of benefits emanating from unit operations, is information on what fields, from an engineering standpoint, are candidates for unit operations and what magnitude of benefits could be expected from each. One hears quite often the claim that *many* fields can be developed and produced efficiently without unitization. No doubt this is true. Others claim that unitization is an absolute necessity to achieve efficiency in *most* fields. Until some factual information is available that provides the basis for estimating costs and benefits of unitization, the arguments concerning its efficacy cannot rise above invective. It does not seem unreasonable to hope that the regulatory authorities in each state could begin to gather such information. Again one encounters the problem of old fields and new fields—and even undiscovered fields.

A Classic Case of Problems Inherent in Voluntary Unitization

The following story, quoted with some omissions, appeared in the Dallas *Morning News* on October 20, 1963:

NEW ORLEANS, LA.—Two Dallas engineers told the Society of Petroleum Engineers meeting here Wednesday that unitization plans for the giant Fairway field in East Texas are all but wrapped up on the operator side, and the concluding step will come shortly with submission of basic agreements to royalty interests.

J. R. Latimer, Jr., chief reservoir engineer for Hunt Oil Co., the selected operator for the unit, and Fred L. Oliver, president of Oliver & West, Inc., consulting organization, described the $2\frac{1}{2}$-year preparatory program as an example of "major cooperative efforts to unitize a large oil field within three years after discovery so that pressure maintenance may be started to achieve maximum recovery.... Final (unitization) papers are now ready to be circulated among more than 150 working interest owners," Latimer and Oliver reported. They added installation of the pressure program to follow acceptance of the unitization agreement by working and royalty interests "may increase ultimate recovery (at Fairway) from around 70,000,000 barrels to about 200,000,000 barrels."

The engineers traced Fairway's development from discovery to the present and pointed out that a first major step toward eventual unitization came when the Railroad Commission approved a well-spacing program which permitted 160-acre development for the James lime reef reservoir. They said of 146 wells producing from this pay, only two are on 80 acres or less.

They said the spacing program allowed early definition of the field reservoir at greatly reduced cost. Next was construction of a gas processing plant to conserve that product and its liquids and set the stage for reinjection of gas into the reservoir for pressure maintenance upon final approval of unitization....

Latimer and Oliver said a major hurdle was passed when, "after long negotiation and many meetings," the unit participation formula was approved by some 90 per cent of the field ownership. It provides for participation based half on acreage and half on acre-feet of pay to Jan. 1, 1976, or the production of 65,000,000 barrels (whichever is first), after which participation will be entirely on acre-feet. They noted that last March the rail commission set the field allowable half on acreage and half on acre-feet.

Said Latimer and Oliver: "Fairway is a significant example of the value and importance of early recognition of reservoir properties and forthright action to achieve maximum recovery without needless cost. It is an example of using technology to make the most of our resources, and should serve as a guide to more effective utilization of our future discoveries. It shows plainly that efficient recovery can be achieved under existing state regulation."

Because of the inability of ownership interests to agree on an allocation formula for the Fairway Field, the voluntary unitization plans were stalled and not put into effect until October 1, 1965, 5 years after the field was discovered and attempts at unit operations were begun. In February, 1965, the Texas Railroad Commission, viewing the declining reservoir pressure in the field with some alarm because of the irreparable damage it could do to the contemplated miscible displacement plan, reduced the allowable for wells in the field to a 5 per cent market demand factor as compared with about a 28 per cent market-demand factor for the state generally at that time. The effect was to reduce field production from about 12,000 barrels daily to 2,000 barrels daily.[24] This appears to have been an obvious attempt on the part of the commission to bring about unitization by means of production controls and in the absence of a compulsory unitization statute. In reviewing the history of the Fairway Field in December, 1965, the *Oil and Gas Journal* noted: "With the drastic cutback in production, there is still the opportunity for optimum recovery from Fairway, and a chance to unitize a big, complex, varied-ownership field in the early years of its life."[25] Clearly, production control is a cumbersome and inefficient solution to this type of problem, and one which gives short shrift to protection of property interests, since production penalties are imposed on everyone, regardless of willingness to cooperate. The Fairway Field, long touted as a classic example of good conservation techniques, came

---

[24] *Oil and Gas Journal*, February 15, 1965, p. 74.
[25] December 6, 1965, p. 118.

painfully close to becoming a classic example of the physical and economic waste that can result when an effort at good field development and production practices is crippled by inadequate conservation statutes and regulations. It should be added that had the agreement not been put together by December 31, 1965, the existing agreement would have fallen apart and attempts to get a new agreement begun again.[26]

The statement quoted above from Latimer and Oliver that "it shows plainly that efficient recovery can be achieved under existing state regulation" was an unfortunate case of counting chickens before the eggs hatched. One can only ponder the question of how many other opportunities have been, are being, and will be lost to multiply recovery through adequate regulations that require efficient reservoir development and depletion.

STANDARDS IN THE ABSENCE OF UNIT OPERATION

Short of unit operation, there are a number of things that can be done to increase ultimate recovery. To start with, we may quote loosely from Buckley's *Petroleum Conservation* to the effect that, for relatively effective recovery, some pools require only (1) well-spacing rules, (2) control of production, and (3) restriction of loss of driving force.[27] This was said as an exception to the more general rule that the entire reservoir should be operated as a single producing unit. The actual practices of conservation now widely utilized in many states turn the exception into the rule, and depend upon the three requirements mentioned.

The Individual Well

There are certain things that can be done at the level of the individual well to prevent dissipation of reservoir energy, such as proper well completions, prevention of venting or flaring of gas, and the prescription of maximum allowed gas-oil ratios. These can have a substantial effect on ultimate recovery whether a field has been unitized or has been developed with unrestricted drilling and

[26] *Oil and Gas Journal,* August 2, 1965, p. 73.
[27] Buckley, *op. cit.,* p. 271.

otherwise unrestricted production. Most states have fairly stringent rules on these points.

Control of Production

A regulated rate of production from the reservoir as a whole, designed to preserve the efficiency of the dominant driving mechanisms, is also important to ultimate recovery. Again, this is true whether there has been unitization or not. This necessity would exist quite apart from any other reasons for restricting production, such as market stabilization. The concept of MER, or maximum efficient rate of production, expresses this factor. (As a technical and economic factor, this will be dealt with elsewhere in this study; see Chapter 7.) For practical purposes of setting rates of production from reservoirs, MER appears at present to have no significance, except in California, since restriction to market demand in most instances gives a lower level than MER. It may, however, acquire significance in states like Wyoming that do not allocate to market demand, and it may be used even in market-demand states as a check against inefficient rates of withdrawal in some fields under existing proration formulas.

Going back to the origins of conservation measures, we have argued that one reason, if not the primary one, for restricting production was to stabilize markets and this, we feel, has remained the primary reason. The aggregate production permitted for stabilization purposes need have no relation to the aggregate amount arrived at by limiting reservoirs to some defined MER. However, since the former amount is typically smaller than the latter would be, it serves the purpose of protecting natural reservoir drives in the usual case. There is in this narrow sense thus no conflict between the market stabilization and the physical recovery purposes.

Well Spacing

The spacing of wells is a third factor that may have an important bearing on efficient recovery. No easy generalizations are possible on this subject, because much depends upon the peculiar characteristics of particular reservoirs. In some cases, it appears that the number of wells is not an important factor for ultimate recovery and, in other cases, the contrary. In any case, however, the number of wells is an important economic factor. If fewer wells will serve to drain the reservoir efficiently at the desired rate, a large number represents economic waste from investment in unnecessary wells. Also, spacing may be an important determinant of productive capacity and, to the extent that there is excess capacity, close spacing creates additional disincentives to exploratory drilling.

The regulation of well spacing involves complicated considerations that are only in part related to efficient recovery. The elements of the problems are made up of (1) efficient drainage subject to the economic factor of the desired rate of withdrawal, (2) the interests of small-tract owners, particularly with respect to drilling, (3) the property rights of tract owners, and (4) the investment costs of development.[28] All of these elements are bound up with the formulas used by regulatory agencies for assigning production allowables to producing wells. The problems of well spacing, therefore, lie inside the most controversial areas of regulatory practice, where the conflicting interests of various segments of the industry come into focus.

The well-spacing provision of the IOCC model statute is designed to cover both equitable and efficiency aspects. The correlative rights rule is to "afford a reasonable opportunity to each person entitled thereto to recover or receive the oil or gas in his tract or tracts or the equivalent thereof without being required to drill unnecessary wells or to incur other unnecessary expense." The well-spacing rule is that "the size of the spacing units shall not be smaller than the maximum area that can be efficiently and economically drained by one well."[29] The latter rule is designed to avoid unnecessary investment. Given this situation, tract owners are to share according to some technical assessment of the respective amounts of recoverable oil underlying their respective tracts.

Application of the spacing rule would permit more economical development of reservoirs than in the absence of such a rule, but it might nevertheless result in much less economical development

---

[28] For an interesting discussion of the economic interrelationships between well spacing and efficient recovery over time, see J. W. McKie and S. L. McDonald, "Petroleum Conservation in Theory and Practice," *Quarterly Journal of Economics*, Vol. 76 (February, 1962), pp. 98–121.

[29] The Governors' Committee report states that, "Since time and economics cannot be ignored, the optimum spacing is concluded to be the maximum number of reservoir acres that would be economically and effectively drained by one well within a reasonable period of time." IOCC Governors' Special Study Committee, *op. cit.*, p. 57.

than would unit operation. In the words of the 1964 IOCC conservation study, "Under unit operation, freedom to locate wells in con-
formance with the structural characteristics of the reservoir and to
utilize fully the reservoir-drive mechanism will permit more efficient
recovery with fewer wells."[30] A spacing rule places some constraints
on where wells can be located on the producing structure. No
doubt, in some situations, a spacing rule would not hamper the
selection of optimum well sites; in others, it would. In the absence
of unit operations, there appears to be technical agreement that,
in the usual case, the best approach to reservoir development is
wide spacing at the beginning to establish the characteristics and
boundaries and then determination of a plan of development based
on the information acquired. Economic considerations could then
control the infill drilling according to the desired rate of withdrawal.
No state has specifically organized its spacing rules on this principle, but something of the sort appears to result from order No.
29–H issued by the Louisiana Commissioner in 1960, which pre-
scribes a 40-acre spacing formula for the first two wells, with later
determination of the general spacing pattern. The IOCC Governors'
Committee endorsed this principle in 1964.[31]

Generally speaking, the economic advantage of owners and operators, particularly if relatively stable prices are assumed, lies on the
side of fairly wide spacing, since it reduces both capital and operating costs. But progress in this direction has been retarded in some
states by what are at bottom mainly political pressures that emanate
from small-tract owners, the drilling contractor industry, and local
communities with an interest in oil-field payrolls. The allowable
formulas being what they are,[32] incentives are also provided, even
to larger operators, to favor relatively dense drilling, since it permits
a more rapid payout of invested capital. The advantage lies not in
denser drilling as such, but in the formulas used for assigning
allowables. The subject will therefore necessarily arise at a later
point where we are dealing with the proration system. We should
note at this point, however, that the introduction of the 1965 Yardstick in Texas removes some of the incentives for dense spacing in
that state. This was an important step in the right direction.

---

[30] *Ibid.*, p. 60.
[31] *Ibid.*, p. xvii.
[32] These are described in Chapter 6.

Some part of the opposition to wide spacing also appears to arise, not from any explicit economic advantage, but from the same habit of mind that opposes unit operation—namely, the idea of a "free competitive system" that permits the individual operator to engage in an enterprise of his own wherever he finds the opportunity. Large drilling units must be managed by a single operator and reduce the active participation of all owners of separate tracts. This may lead to uncertainties and suspicions among the participants.

To the extent that states may be inclined to move toward unit operation, the enforcement of wide well spacing is probably a necessary intermediate step. It breaks down the separatist attitudes that support excessive investment in development, it provides experience in cooperative action, and it demonstrates the economic advantages to owners. This could lead over into the acceptance of cooperative action on a wider scale, in view of the increasingly obvious financial advantage both to many of the affected interests and, in the long run, to the economic base and fiscal support of the states.

The compelling argument of economic necessity appears to be in process of promoting the use of larger drilling units. By requiring large units and revising its allowables in favor of them, a state can make itself an attractive field for exploratory activity. Given the situation of stable or falling crude oil prices, exploratory expense needs to be backed up by less development expense. Thus, spacing and allowables together become competitive weapons in the hands of particular states to attract exploration. With the existence of unused producing capacity and little chance for significant price increases, the logic of this has existed all along. But it becomes more compelling as the proration system tends to work itself into an impasse by driving down rates of return on unnecessarily large investment outlays. It appears that the strong pressure for wider spacing has recently come from the companies themselves, large and small, rather than being generated within the regulatory agencies. Compulsory action, however, will be required to drive minority interests into line, and it will take revised systems of allowables to make wider spacing more attractive.

There is, unfortunately, only sparsely scattered information which shows the distribution of wells according to a size scale. The 1964 IOCC study contains a tabulation, reproduced in Table 1, that shows oil-spacing orders issued by states according to the size of

TABLE 1. REPORTED OIL SPACING ORDERS ISSUED BY STATES, 1940–62

States	20 Acres or less						40 Acres						80 Acres						160 Acres or more					
	1940	1950	1955	1960	1961	1962	1940	1950	1955	1960	1961	1962	1940	1950	1955	1960	1961	1962	1940	1950	1955	1960	1961	1962
Alabama	0	0	0	0	0	0	0	1	0	2	0	0	0	0	0	2	1	1	0	0	0	4	1	0
Alaska	–	–	–	–	–	–	–	–	–	–	–	0	–	–	–	1	0	0	–	–	–	0	0	0
Arkansas	1	2	7	2	6	2	1	1	6	1	0	1	0	0	1	0	0	2	1	1	1	2	0	0
Colorado	–	–	0	0	0	0	–	–	23	0	0	1	–	–	0	0	2	0	–	1	0	0	0	0
Illinois[a]	–	–	–	–	–	–	–	–	–	–	–	3	–	–	–	–	–	1	–	–	–	–	–	–
Kansas	–	–	–	3	2	2	–	–	–	30	35	33	–	–	–	3	19	15	–	–	–	–	–	–
Louisiana	2	2	4	10	23	8	3	20	10	23	26	17	0	7	6	12	39	40	0	0	0	3	0	5
Michigan	3	8	0	4	5	1	0	5	6	6	9	40	0	0	0	0	0	0	0	0	0	0	0	0
Montana	–	–	1	1	0	3	–	–	0	1	0	4	–	–	10	1	1	1	–	–	1	1	2	4
Nebraska	–	0	0	0	0	0	–	–	–	–	1	3	0	0	0	1	1	0	–	–	–	–	0	0
New Mexico[b]	–	–	–	–	–	–	–	–	1	–	–	–	–	1	–	10	2	12	–	0	0	0	0	4
North Dakota	–	–	0	0	0	–	–	–	–	–	–	–	–	–	–	3	2	–	–	–	–	–	–	–
Oklahoma	19	21	115	42	69	57	2	18	37	53	110	90	0	0	2	41	94	104	0	0	0	6	0	3
Texas	–	30	31	14	8	5	–	24	44	69	76	55	–	–	19	22	37	38	–	–	–	0	7	6
Utah	–	–	–	0	0	0	–	–	–	0	0	0	–	–	1	1	–	0	–	–	–	0	4	0
Wyoming	–	–	1	0	0	0	0	1	9	0	1	1	0	8	4	8	5	8	0	1	2	0	0	1
Totals	25	63	159	76	113	78	6	70	136	185	259	248	0	8	46	106	207	222	1	4	16	15	23

[a] Authority for spacing orders first granted by 1960 legislature.

[b] Statewide 40-acre minimum spacing.

NOTE: Florida has 40-acre spacing by statute; all of Indiana's spacing by statute is 20 acres or less.

SOURCE: IOCC Governors' Special Study Committee, A Study of Conservation of Oil and Gas in the United States (Oklahoma City, 1864), p. 59.

the drilling unit. The trend toward larger drilling units is quite apparent and will no doubt continue. Unfortunately, the IOCC made no effort to record the size of the fields in terms of reserves or production in each size category of drilling units. Thus, it is impossible to measure how significant the trend to larger units is in terms of output.

A certain amount is known about the differing situations in various states. Since the pre-1930 days of close spacing in the Artesia Area, New Mexico has had (since 1935) relatively large drilling units and has been under no strong pressure to introduce small-tract drilling. Precisely the opposite has been true in Texas in the past, though apparently a change is due. In 1963, Texas established a state-wide spacing rule of 40 acres that can be modified by special field rules. Louisiana has recently taken a long step toward larger units. Within single states, much of the development has proceeded by special field orders rather than by general rule, and some states, like Oklahoma and Kansas, have no general rule at all. The whole subject is therefore a no-man's-land of ignorance insofar as gaining a clear picture of well spacing as it is practiced. As far as we know, no regulatory agency in any important producing state provides a statistical analysis of the spacing situation in the state.

A state's intent to promote high recovery at minimum cost, short of unit operation, is tested by the degree to which sound well-spacing rules are enforced from the beginning of the development of new reservoirs. Some states have gone much further than others in this direction, but we know of no state that has set up its rules in away to apply the test fully. North Dakota possibly leads the way in incentives for wide spacing. By its action in 1960, Louisiana took a step forward. Since economy in development seems to be the wave of the future, as the necessary means to continuing profitability in many situations, presumably this is an area that will be subject to reexamination in many states.

APPENDIX TO CHAPTER 4

WELL SPACING AND PRODUCTION CONTROLS:
AN ECONOMIC MODEL

We have noted at several points in this chapter that the economic decision made by a company for spacing wells is a function of investment in the wells and the revenue realized from the wells over time. Revenue, of course, is a function of the price per barrel and the output per unit of time. Output at any point in time depends on three things: (1) the natural reservoir characteristics, (2) the effect of past output (or looking forward, the impact of current output on future output), and (3) the allowable set by the regulatory commission (assuming a market-demand state).

The following discussion describes a simple model of oil-well drilling and production over time and shows in an abstract fashion the effects of regulations on the production process.[33] Assume for the time being that we have an oil reservoir of known size and characteristics that has been discovered by the drilling of an exploratory well. Assume also that a single operator controls all the acreage and that there are no well-spacing or production regulations. Assume also competitive crude oil and factor markets and a going price for oil that will not be influenced by production from this new field. We assume that this operator wishes to develop and produce this field so that he maximizes his net revenue over the life of the field. Three characteristics of the field must also be noted. First, beyond some critical rate of production, ultimate recovery from the reservoir decreases as the rate of production increases. Second, ultimate recovery is independent of the number of wells in the reservoir, but the recoverable oil can be produced more rapidly with more wells. Third, average and marginal lifting costs rise as the rate of production increases.

Given these assumptions, the goal of profit maximization over time is realized if

$$MR_o - MC_o = (MR_t - MC_t)\left(\frac{1}{(1+i)^t}\right)(1 + r_t)$$

where: MR represents long-run marginal revenue
MC represents long-run marginal costs, including marginal operating costs and marginal well costs

---

[33] This model is adapted from one developed by Stephen L. McDonald. See his paper, "Percentage Depletion, Expensing of Intangibles and Petroleum Conservation," before the Conference on Tax Treatment of Exhaustible Resources, University of Wisconsin, Milwaukee, August 18, 1964, to be published in Mason Gaffney (ed.), *Conservation, Competition and Taxation* (Madison: University of Wisconsin Press).

> i is the rate of interest for the operator, adjusted for risk and uncertainty
>
> o is the subscript indicating the present
>
> t is the subscript indicating any point in future time
>
> r_t is the fraction of a barrel lost from ultimate recovery, and which might have been recovered in time t, for every barrel of production transferred from time t to time o.

Let us for a moment leave the $(1 + r_t)$ expression out of the equation, which means that the ultimate recovery from the reservoir is not rate sensitive, i.e., regardless of what rate of production we pick today, it will have no effect on ultimate recovery. We then have the equation:

$$MR_o - MC_o = (MR_t - MC_t)\left(\frac{1}{(1 + i)}t\right).$$

If we further assume a zero rate of interest $(i = 0)$, the second term on the right-hand side of the equation (the discount term) drops out. We would then be saying that we adjust today's rate of production (in time o) and tomorrow's rate of production (in time t) so that our net revenue from the last (marginal) unit produced today, $MR_o - MC_o$, equals the net revenue from the last unit produced tomorrow, $MR_t - MC_t$.[34] If we assume that prices will remain the same $(MR_o = MR_t = \text{constant})$, then the only variables are the costs of the last units today and tomorrow. One of our initial assumptions was that marginal costs rise as the rate of production increases. If $MC_o > MC_t$ under our limiting assumptions, it would pay to reduce the rate of production today and/or increase the rate of production tomorrow, since this would lower MC_o and/or raise MC_t. What specifically would be done will depend on the relationships of MC_o and MC_t to price $(MR_o = MR_t)$. For example if $MC_o > MR_o < MC_t$, the producer can add to his net income stream by reducing production today until $MC_o = MR_o$ and by increasing production tomorrow until $MC_t = MR_t$. The excessive rate of production today means that on the marginal barrel today the producer is losing money, therefore his net income rises if he backs off until $MC_o = MR_o$. Tomorrow's rate of output,

---

[34] It is important to keep in mind that we are always working with marginal or incremental barrels of production and not total production.

for which $MC_t < MR_t$, is too slow since the producer could add to his net income by producing a few more barrels. He raises tomorrow's rate until $MC_t = MR_t$. He then adds to income both today and tomorrow by adjusting production rates in the manner described.[35] We thus have maximized income for the combined periods of today and tomorrow when:

$$MR_o - MC_o = MR_t - MC_t.$$

Let us now look at the situation with a discount factor taken into account. Since a dollar of income today is worth more than a dollar of income tomorrow, we must discount tomorrow's income to get its value today, i.e., to get the present value of future income. The rate at which we discount tomorrow's income is independently determined, and all we can say is that it is some interest rate earned on investment by the oil operator, adjusted for risk and uncertainty. The higher the rate of interest, the lower the present value of future dollars. One might conclude then that, other assumptions remaining the same, the higher the interest rate the faster the rate of output today and the slower the rate of output tomorrow. This gives us the equation:

$$MR_o - MC_o = (MR_t - MC_t)\left(\frac{1}{(1+i)}t\right).$$

The last term in the equation $(1 + r_t)$ is unique to oil production, and reflects the fact that high rates of production today can actually reduce the total amount of oil that can ultimately be recovered. The term r_t is that fraction of a barrel of ultimate recovery that is lost by producing faster than some optimum physical rate today. Put in other terms, r_t is the fraction of a barrel of oil which might have been recovered tomorrow (time t) for every barrel of oil produced today instead of tomorrow at a rate in excess of some MER. It follows then that in a physical sense you may not be able to swap a barrel of oil today for a barrel of oil tomorrow. A barrel of oil produced today forces you to give up more than a barrel of oil produced tomorrow. You sacrifice $1 + r$ barrels of tomorrow's oil to produce 1 barrel today. It is obvious you may be

---

[35] In this discussion we have dealt only with marginal operating costs on existing wells. An alternative in the long run is to vary the number of wells so that marginal operating costs *plus* marginal well-drilling costs are equated with marginal revenue.

quite willing to make this sort of physical swap because, if we assume any positive interest rate, a barrel of tomorrow's oil is worth less than a barrel of today's oil at any constant price. This brings us back to the profit maximizing equation:

$$MR_o - MC_o = (MR_t - MC_t)\left(\frac{1}{(1+i)}t\right)(1 + r_t).$$

We can relax our assumption about constant prices without affecting the analysis. Higher expected future prices would tend to push the production rate up in the future and down today, other things being as assumed.

The model has assumed competitive conditions so that the maximizing conditions for the operator represented in the equation are also the conditions that maximize benefits to society as a whole.[36] In other words, there are no costs that are external to the operator and that must be borne by society. This assumes away such things as possible pollution of ground- or surface-water supplies and possible damage to land, vegetation, etc. If these occurred, as they might well in a case of salt water disposal, there would be costs to society not borne by the operator. Thus we may say that given single ownership of reservoirs, competitive markets, and no costs external to the operator, our equation represents optimum conservation policy from society's view. The rates of production over time dictated by the model give the operator the maximum stream of net income from the resource over the life of the resource and also give society the greatest net benefits from the resource over its life. In other words, these rates of production are ideal from a conservation standpoint.

If we relax the assumption about single ownership of the reservoir, we introduce another term into the equation. Let us assume several operators drilling in and producing from a common source of supply. Let x represent the fraction of a barrel of oil lost by one of the operators to a competing operator in time t for every barrel of oil not produced by the first operator today, time o. In other words, x is the loss to one operator from drainage by another operator. The equation for each operator under these circumstances

---

[36] This statement assumes that the operator's discount rate is in fact the same as the social discount rate, adjusted for the risk and uncertainty of the particular activity.

then takes the form:

$$MR_o - MC_o = (MR_t - MC_t)\left(\frac{1}{(1+i)^t}\right)(1 + r_t)\,(1 - x_t).$$

The effect of competitive production that yields a high x will be for each operator to drill and produce as rapidly as possible today. Failure by one operator to do so means that his net income stream over the life of his portion of the resource is less. The obvious effect for the reservoir as a whole is to reduce ultimate recovery by increasing the number of wells drilled and by increasing the rate of production of each well in time o.

Throughout the discussion we have lumped operating costs and well-drilling costs together in a long-run marginal cost concept. This is done to emphasize the relationship between the number of wells and the rate at which a reservoir is produced. Optimal conservation policy dictates that both the optimum number of wells and the optimum rate of production be determined simultaneously.

In summary then, an oil operator practices conservation in the economist's view by drilling the optimum number of wells and producing this number at the optimum rate of production, these optima being where the present value of the reservoir is at a maximum. Present value is maximized where the discounted marginal net profit in each period is equal to that in every other period, which means that the marginal rate of profit is equal to the current rate of interest.

Let us now introduce several types of regulation and consider some possible results. The imposition of MER controls, i.e., not allowing a reservoir to be produced at a faster rate than will yield maximum ultimate recovery, constrains the r term in the equation to zero. In other words, no operator can trade a barrel of oil today for more than a barrel tomorrow. This may or may not be beneficial to society. If the MER is quite low, society might be better off sacrificing less valuable barrels tomorrow for more valuable barrels today.[37]

Spacing regulations attempt to constrain the x term to zero but do not do so very systematically. Under any given spacing pattern assigned to a field with competing operators, it may be possible for a single operator to drill wells up to the density allowed and

---

[37] The MER concept is not at all clear in industry usage. See Chapter 7.

drain oil from neighboring property. This is true even with production controls. Because of this drainage aspect, there are clearly some incentives to overdrill if the optimally efficient pattern is less dense than the pattern prescribed by the regulation. Unitization is perhaps the only way of assuring that x is zero and that overdrilling is avoided.[38]

Turning to a related aspect, let us assume that we have production allocation by a state so that allowed production is a function of the number of wells in the field, i.e., so many barrels per well. The only way that more revenue can be gained from the field is by incurring additional costs in drilling more wells. Assume no spacing controls or an imposed spacing pattern in which the minimum acreage per well is significantly less than optimum spacing. The alternative of adjusting the producing rates of wells to maximize profits is no longer available. Thus, while it may be profitable for the operator to put down more wells (up to the point where the last well costs him as much as it returns him), it is not as profitable to him as would be the case of allowing discretion in producing rates as well. This would even be true if allocations were made on an acreage basis, although it is likely that drilling would be less dense under these circumstances.

If we impose well spacing and market-demand proration with allowables based on some depth-acreage schedule,[39] the picture gets very confused. Both paths of action available to operators for profit maximization in an unregulated situation are now restricted to some degree. Given the limit on production imposed by the yardstick and market-demand factor, the operator can increase profits only by drilling wells up to the limit allowed by the spacing regulations. In each case it comes back to the logic of the model—marginal costs are equated with marginal revenues produced. However, with regulation, the parameters within which the operator makes his decisions are drastically altered so that his decision is economically rational for him, given these constraints, but the constraints themselves may force the decisions to be uneconomic and irrational from society's standpoint. In other words, profit maximization by

---

[38] Even with unitization it is possible to have overdrilling if a field allowable is based on a depth-acreage schedule similar to those currently in effect in the market-demand states. See Chapter 6.

[39] See Chapter 6 for a discussion of depth-acreage schedules.

producers will not yield optimum conservation for society. The MC_o, MR_o, MC_t, and MR_t facing the producer are heavily influenced by the regulations imposed.

Perhaps we have carried the analysis of the simple model far enough to demonstrate our two main points: (1) that well spacing and production rates are inextricably interwoven and interdependent so that they must be considered simultaneously in making judgments about economically efficient development and depletion of an oil reservoir; and (2) the imposition of a given set of regulations on well spacing and production rates creates an entirely different framework for analyzing the path to profit maximization for the operator than exists if no regulations are imposed or if a different set is applied.

We intentionally have not introduced such further complicating factors as uncertainties of future reservoir performance when viewed early in its life, uncertainties about the technical feasibility of alternative pressure maintenance or secondary recovery operations, and uncertainties with respect to future costs and prices. These are real problems and cannot be ignored. They do not, however, add to understanding the problems we have tried to outline here.

Chapter 5

Excess Producing Capacity and Related Problems

The existing system of state conservation regulation operates within a situation in which there is a large oversupply of oil or, put more precisely, excess oil-producing capacity in the field relative to demand. This situation, on the one hand, creates serious problems for specific groups within the industry and for the regulatory authorities, and places a severe strain upon the system. On the other hand, the regulatory practices are themselves responsible in considerable degree for the existence of the situation. In addition to domestic oversupply, there also exists a situation of excess producing capacity outside the United States relative to the demand in world markets—a fact which adds further difficulties to control of the domestic market. It therefore is necessary at this point to examine in more detail (1) what is meant by excess producing capacity, (2) where it is located geographically in the nation, (3) what has caused the excess, (4) whether it is a short-run or long-run phenomenon, and (5) what are some of the conservation problems caused by it.

MEASUREMENT AND LOCATION OF PRODUCING CAPACITY

Estimates of crude-oil-producing capacity for the United States are made by two groups: the National Petroleum Council (NPC) and the Independent Petroleum Association of America (IPAA). It is impossible to determine specifically what, if any, interdependence there is between the two estimates. The NPC has made estimates as of January 1, 1947, 1951, 1953, July 1, 1954, January 1, 1957, 1960, and 1964. The IPAA has made estimates annually since 1954. In addition, the Petroleum Administration for War (PAW) made estimates during World War II, and the Petroleum Adminis-

97

tration for Defense (PAD) made estimates during the Korean Crisis (1951 and 1952). Both the NPC and IPAA report publicly the estimates by PAD districts, and in 1965 the IPAA began reporting on a state basis and, in some cases, a state subdivision basis.[1] The NPC makes state estimates, however, and provides the IPAA with this information. Since the NPC committee, which is responsible for the productive-capacity reports, works closely with the API and American Gas Association (AGA) committees on reserves, it would appear that the NPC is the primary source for productive-capacity data. The IPAA productive-capacity committee does not have direct access to reserve estimates by fields, but does, of course, have the published reserves estimates by states, as well as information on production, drilling activity, producing wells, etc. We will use the NPC data in our discussion primarily, with some references to IPAA and IOCC figures for individual state data; we shall make no attempt to reconcile the three series.

The NPC discusses productive capacity as the rate of production that could be sustained for a brief period without further drilling.[2] There is no discussion of what criteria are used in establishing this rate of production, nor is there a discussion of what is meant by a brief period. Presumably, this is a concept indicating the rate at a given instant of time.[3] The IPAA definition differs in that it considers a rate of production that can be sustained 6 to 12 months without additional drilling and without significant loss of ultimate recovery.[4] The IOCC defines capacity as the "maximum average daily production that could be sustained for a period of six months under existing conservation regulations."[5] Thus it is somewhat different, in terms of time horizon, from both the NPC and IPAA definitions. By definition the IPAA figures for any given time should be somewhat lower than NPC figures. Both groups point out that these estimates are made without regard for transportation,

---

[1] In the spring of 1966, the Energy Resources Committee of IOCC gathered data on crude-oil-producing capacity by states. These figures were reported at the June, 1966, meeting of the IOCC. See *Oil and Gas Journal*, June 27, 1966, p. 97.

[2] NPC, *Proved Discoveries and Productive Capacity* (1961).

[3] By using the language "capacity *on* January 1, 1964" [emphasis supplied], the NPC strengthens this interpretation. See NPC, *Proved Discoveries and Productive Capacity* (1964), pp. 12–15.

[4] IPAA, *Report of the Productive Capacity Committee,* May, 1966.

[5] IOCC, "Memorandum to State Oil and Gas Conservation Agencies from the IOCC Energy Resources Committee," Spring, 1966.

storage, or processing facilities and without regard for market. It should also be stressed that these estimates are made for producing wells as they are currently operating. No attempt is made to estimate changes in capacity that might result from well cleaning or workover, the installation of larger pumping units and other production equipment, or the stimulation of producing formations. Productive capacity, therefore, should not be confused with the concept of availability, which means capacity to produce *and* deliver to required markets.

NPC productive-capacity figures for crude oil exclude lease condensate, a product that is indistinguishable from crude oil in lease tanks because of the mixing of the two. IPAA figures, on the other hand, include both crude oil and lease condensate. This difference in definition reduces the NPC figures relative to those of the IPAA for any given year. The U.S. Bureau of Mines estimated that of total 1963 production of 7,431,000 barrels daily, 7,121,000 barrels were crude production and 310,000 barrels were lease-condensate production.[6] This indicates that 4.2 per cent of production in that year was lease condensate. It is perhaps reasonable to assume that at any point in time lease-condensate-producing capacity stands in at least the same ratio to crude-producing capacity as lease-condensate production does to crude production.[7]

Table 2 shows crude oil production capacity as reported by the NPC for January 1, 1951, 1953, 1957, 1960, and 1961, and July 1, 1964. A production figure is calculated for the same dates by averaging the production rate for the preceding and following years. Production is from the Bureau of Mines and includes lease condensate. It should be remembered, when making comparisons with these production data, that NPC capacity figures exclude lease condensate and thus are understated to that extent. For each year, we have shown the amount of excess capacity, and we also have shown production as a percentage of excess capacity. The purpose of the table is not to show exact amounts of excess capacity,

---

[6] U.S. Department of the Interior, *An Appraisal of the Petroleum Industry of the United States* (Washington: U.S. Government Printing Office, 1965), Table 28. Note to table indicates figures differ slightly from those found in the *Minerals Yearbook.*

[7] For a full discussion of the NPC and IPAA concepts, see W. F. Lovejoy and P. T. Homan, *Methods of Estimating Reserves of Crude Oil, Natural Gas, and Natural Gas Liquids* (Washington: Resources for the Future, Inc., 1965), pp. 38–42.

TABLE 2. COMPARISON OF CRUDE-OIL PRODUCTIVE CAPACITY AND PRODUCTION IN THE UNITED STATES, 1951–64[a]

(thousands of barrels daily)

PAD districts	January 1, 1951[b]				January 1, 1953[b]			
	Productive capacity	Actual production	Excess capacity	Production as % of prod. capacity	Productive capacity	Actual production	Excess capacity	Production as % of prod. capacity
I. East Coast	54	54	0	100.0	49	50	−1	102.0
II. Mid-Continent	1,083	1,077	6	99.4	1,238	1,146	92	92.6
III. Gulf Coast	4,161	3,525	636	84.7	4,686	3,838	848	81.9
IV. Rocky Mountain	350	292	58	83.4	394	332	62	84.3
V. West Coast[c]	1,079	959	120	88.9	1,098	991	107	90.3
Total U.S.	6,727	5,907	820	87.8	7,465	6,357	1,108	85.2

PAD districts	July 1, 1954				January 1, 1957[b]			
	Productive capacity	Actual production	Excess capacity	Production as % of prod. capacity	Productive capacity	Actual production	Excess capacity	Production as % of prod. capacity
I. East Coast	43	43	0	100.0	37	37	0	100.0
II. Mid-Continent	1,380	1,169	211	84.7	1,591	1,355	236	85.2
III. Gulf Coast	5,224	3,728	1,496	71.4	6,613	4,299	2,314	65.0
IV. Rocky Mountain	561	427	134	76.1	615	525	90	85.4
V. West Coast[c]	1,123	975	148	86.8	1,011	945	66	93.5
Total U.S.	8,331	6,342	1,989	76.1	9,867	7,161	2,706	72.6

	January 1, 1960[b]				January 1, 1964[b]			
I. East Coast	29	29	0	100.0	30	30	0	100.0
II. Mid-Continent	1,555	1,334	221	85.8	1,473	1,330	143	90.3
III. Gulf Coast	7,331	4,168	3,163	56.8	8,399	4,711	3,688	56.1
IV. Rocky Mountain	664	672	−8	101.7	678	657	21	96.9
V. West Coast	1,006	848	158	84.3	1,010	853	157	84.4
Total U.S.	10,585	7,051	3,534	66.6	11,590	7,581	4,009	65.4

[a] Production includes lease condensate.
[b] Production on January 1 calculated by averaging immediate past year and current year.
[c] Elk Hills included in West Coast figures.

NOTE: Details may not add to totals because of rounding.
SOURCES: Productive capacity from National Petroleum Council; production from U.S. Bureau of Mines.

but rather to give the general magnitudes of excess capacity, the changes in the excess over the thirteen-year period, and the location of the excess. For the nation as a whole, it is clear that excess capacity has been growing during the decade, although the rate of growth of the excess seems to have slowed down between 1957 and 1964. In 1951, actual production was 87.8 per cent of productive capacity; by 1964, actual production was only 65.4 per cent of productive capacity. While there is no general agreement on how much excess capacity the industry should have in order to allow smooth operation within the industry and to provide for national security, a figure of 10 to 15 per cent excess has been mentioned from time to time in both industry and government circles. Even in PAD District III, the excess in 1951 was not overly large, since production was almost 85 per cent of capacity.[8] In the years following 1951 the situation changed quite rapidly as production grew more slowly than producing capacity.

By 1964, production had fallen to 56.1 per cent of capacity in PAD District III. Because the PAD figures are not reported by states, they do not show how heavily the excess capacity is concentrated in the two states of Texas and Louisiana. To show this, in Table 3 we use the IPAA figures for the five major market-demand states and for the United States total. For comparison we also show the IOCC figures. By the IPAA figures, at the beginning of 1966 Texas accounted for 53 per cent of national excess capacity and Louisiana for 36 per cent. Most of the rest is attributable to California.[9] If the larger NPC figures were available by states, it can be conjectured that the percentage showing would not be greatly different. The addition of the IOCC figures illustrates the state of confusion into which the reporting of capacity figures has fallen.

MARKET FACTORS RELATED TO EXCESS PRODUCTIVE CAPACITY

In the most general terms, the growth and continued existence of substantial excess producing capacity may be attributed to the unresponsiveness of the oil-producing industry to market forces.

---

[8] District III includes the states of Texas, Louisiana, Arkansas, Alabama, Mississippi, and New Mexico.

[9] U.S. Bureau of Mines reported 1965 production for California was 867,564 barrels daily. The IPAA capacity was 1,050,000 barrels daily.

TABLE 3. PRODUCING CAPACITY AND PRODUCTION IN THE
FIVE MAJOR MARKET-DEMAND STATES, JANUARY 1, 1966

(thousands of barrels daily)

State	Production[a] (1965)	Producing Capacity IPAA[b]	Producing Capacity IOCC[c]	Excess Producing Capacity IPAA	Excess Producing Capacity IOCC
Texas	2,757	4,306	4,500	1,549	1,743
Louisiana	1,630	2,685	2,179	1,055	549
Oklahoma	557	630	650	73	93
New Mexico	326	355	350	29	24
Kansas	287	299	300	12	13
Total	5,557	8,275	7,979	2,718	2,422
Total U.S.	7,805	10,745	10,742	2,940	2,667

[a] U.S. Bureau of Mines, *Monthly Petroleum Statement*, March 14, 1966, Table 4.
[b] IPAA, *Report of the Productive Capacity Committee*, Table II, May, 1966.
[c] *Oil and Gas Journal*, June 27, 1966, p. 97.

More specifically, after the rapid growth of market demand after the war, conducive to expansion, came to an end, the expansive activities continued—at a diminished rate, it is true, but still at a rate sufficient to keep on adding to excess production capacity. To a degree, this may be attributed to the momentum of plans based in expectations of market growth which failed to materialize. But the size and long continuance of the phenomenon make this an inadequate explanation. An important part of the answer, we judge, lies in the insulation against market forces provided by regulation. The way in which regulatory practices directly induce overcapacity will be taken up at later points. At this point a brief examination of demand-and-supply factors will shed some light on the forces at work.

Table 4 shows a breakdown of the components of supply for 1951, 1953, 1955, 1957, and 1959–64. It will be noted that of the new supply in 1951 of 2,761 million barrels of liquid hydrocarbons, domestic crude accounted for 81.4 per cent, domestic natural gas liquids (NGL) accounted for 7.4 per cent, imports of crude oil accounted for 6.5 per cent, and imports of refined products for 4.7 per cent. By 1964, the total supply had risen to 4,036 million barrels and the percentages contributed by the four sources of supply were: domestic crude 69.0 per cent, domestic NGL 10.5 per

TABLE 4. NEW SUPPLY OF OIL IN THE UNITED STATES, 1951-64

Item	1951 Millions of barrels	1951 Per cent of total new supply	1953 Millions of barrels	1953 Per cent of total new supply	1955 Millions of barrels	1955 Per cent of total new supply	1957 Millions of barrels	1957 Per cent of total new supply	1959 Millions of barrels	1959 Per cent of total new supply
Domestic production:										
Crude oil[a]	2,248	81.4	2,357	79.3	2,484	77.1	2,617	76.2	2,575	72.6
Natural gas liquids[b]	205	7.4	239	8.0	282	8.8	293	8.5	321	9.1
Imports:										
Crude oil	179	6.5	236	7.9	285	8.8	342	10.0	352	9.9
Refined products	129	4.7	141	4.7	170	5.3	184	5.3	297	8.4
Total new supply	2,761	100.0	2,973	100.0	3,221	100.0	3,436	100.0	3,545	100.0

Item	1960 Millions of barrels	1960 Per cent of total new supply	1961 Millions of barrels	1961 Per cent of total new supply	1962 Millions of barrels	1962 Per cent of total new supply	1963 Millions of barrels	1963 Per cent of total new supply	1964 Millions of barrels	1964 Per cent of total new supply
Domestic production:										
Crude oil[a]	2,575	71.9	2,622	71.2	2,676	70.2	2,753	70.0	2,787	69.0
Natural gas liquids[b]	340	9.5	362	9.8	373	9.8	401	10.2	422	10.5
Imports:										
Crude oil	372	10.4	382	10.4	411	10.8	413	10.5	439	10.9
Refined products	293	8.2	318	8.6	349	9.2	365	9.3	388	9.6
Total new supply	3,580	100.0	3,684	100.0	3,809	100.0	3,932	100.0	4,036	100.0

[a] Includes lease condensate.
[b] Includes small amounts of benzol.
SOURCE: U.S. Bureau of Mines.

cent, crude-oil imports 10.9 per cent, and product imports 9.6 per cent. It might be argued that imports have supplied a part of domestic demand that should have been filled with available domestic producing capacity, and thus, that imports have caused our excess capacity. Or it might be argued that the growth of domestic NGL production has taken markets that should have been supplied by domestic crude oil, and thus, that NGL's have caused or contributed greatly to excess capacity. Clearly, these two sources of supply have had an impact on domestic crude supplies, but other factors have also been at work. NGL and imports present different problems, and the growth of each has unique aspects that will be discussed below. Before doing this, we must continue our catalogue of factors causing excess productive capacity.

Turning to the demand side we find that the growth in annual crude oil consumption (domestic and imports) between 1951 and 1964 was about 36 per cent. Total energy consumption between 1951 and 1964 grew about 41 per cent, indicating that crude-oil consumption has not kept pace with the total energy market.

Table 5 indicates the percentage contribution of major energy sources to total domestic energy consumption in the United States from 1951 through 1964. During this period, as the table shows, there was a radical shifting of patterns of energy consumption. The most striking facts are the decline of coal from 35.8 to 22.5 per cent of the total, and the rise of natural gas from 19.6 to 30.1 per cent. The decline of coal's share of the market reflects the substitution for coal of both natural gas and fuel oils. This has been a plus factor in keeping oil consumption up. The spectacular increases in the share of natural gas has been at the expense of both coal and oil. Thus, in one sense, the petroleum industry has been its own worst enemy, since natural gas competes directly with the crude-oil-derived heating oils and industrial fuel oils in many areas of the country.

Other elements detrimental to oil show up in Table 5. There was a substantial increase in the share of natural gas liquids. The share of imported crude oil also rose, though this rise was small after the policy of restriction was initiated in 1955 and put on a mandatory basis in 1959. Imported refined products have steadily increased their share of the energy market since 1951, but much of this increase has not been directly competitive with domestic crude oil. Of total product imports in 1963, 76 per cent were residual fuel oil, the increase of residual fuel imports was largely at the expense of coal in markets along the East Coast, not at the expense of domestic oil. Refineries had in fact greatly reduced the production of residual fuel oil, because of its very low price as compared with other products which improved technology permitted them to recover.

TABLE 5.   CONTRIBUTIONS TO U.S. ENERGY CONSUMPTION
BY ENERGY CATEGORIES, 1951–64

(*per cent*)

Year	All coals[a]	Crude oil	Do-mestic crude oil[b]	Im-ported crude oil[b]	Im-ported refined petro-leum prod-ucts[c]	Nat-ural gas– dry	Nat-ural gas– liquids	Water power	Nu-clear power
1951	35.8	37.6	34.8	2.8	0.3	19.6	2.4	4.3	—
1953	31.6	39.5	35.9	3.6	0.5	21.6	2.7	4.1	—
1955	29.3	39.9	35.8	4.1	0.9	23.1	3.0	3.8	—
1957	27.1	40.5	35.8	4.7	0.9	24.8	3.0	3.7	—
1959[d]	23.1	39.1	34.4	4.7	3.2	27.6	3.1	3.9	—
1960	23.2	38.2	33.4	4.8	3.2	28.3	3.2	3.9	0.0
1961	22.4	38.0	33.2	4.8	3.5	28.9	3.3	3.9	0.1
1962	22.0	37.3	32.3	5.0	3.8	29.3	3.4	4.1	0.1
1963	22.3	37.0	31.5	5.6	3.6	29.8	3.4	3.8	0.1
1964	22.5	36.5	31.5	5.0	3.8	30.1	3.4	3.6	0.1

[a] Bituminous, lignite, and anthracite.
[b] Total crude oil broken down by the ratio of domestic crude production to imported crude oil.
[c] Net of exports.
[d] Beginning in 1959, data on a 50-state basis.
SOURCE: U.S. Bureau of Mines.

Interfuel competition is exceedingly complex and the above comments are not intended to show all of the relationships and movements in energy markets. They do, however, furnish another piece in the picture of the causes of excess crude-oil-producing capacity in the United States. In 1951, forecasters of growth rates of potential crude-oil markets were correct in predicting substantial increases. What they did not forecast was the extent to which substitute fuels would capture parts of these markets.

One of the most troublesome aspects of the overcapacity problem to analyze is the continued growth of domestic-oil-producing capacity in the face of some of the forces described above. Table 6 presents data for selected years since 1951 on the total number of wells drilled, oil wells completed, service wells completed, exploratory wells drilled, the number of producing oil wells at the end of each year, and abandonments of oil wells during each year. Drilling hit a peak in 1956 when over 58,000 wells were drilled. It dropped rather sharply to the 43,000- to 46,000-well range and apparently has stabilized. One might have expected that in the face of the 1-million-barrel increase in *excess* productive capacity between July 1, 1954, and January 1, 1957, drilling would have been curtailed to a greater extent than it was. Even at current levels of drilling, the industry seems to be adding somewhat to excess productive capacity, although this is difficult to determine. In particular, however, the industry is not reducing it significantly. The NPC data shown in Table 2 seem to indicate that additions to productive capacity since 1960 have just about offset the increases in demand, so that the industry is no longer adding significantly to the excess in percentage terms. In terms of barrels, there were 447,000 barrels of daily capacity added to the excess in this period (1960–64). The surge in demand in late 1965 and early 1966 resulted in expanded production, and no doubt a concurrent decline in excess capacity. It is too soon to tell at this writing whether or not the increased production rates will be sustained and, if so, what effect if any this will have on the drilling of new wells and the development of new capacity.

The disincentives to drilling new wells that are created by excess capacity and that are felt through reduced producing levels per prorated well in market-demand states have not been particularly effective. The pattern of exploratory drilling follows along the same lines as total drilling, hitting a peak in 1956 of over 16,000 wells and dropping off to the 10,500 to 11,500 range. Part of the reason for the increase in productive capacity may be found in the figures on service wells. During the middle 1950's when the drilling of oil, gas, and condensate wells was at a peak, service well drilling was relatively small. Then when the general drilling trend turned down, service well drilling picked up. One possible interpretation of this might be that oil operators were obtaining new reserves through the development of pressure-maintenance and secondary-recovery projects rather than through as much exploratory or even developmental drilling. A reason for this route to new reserves

TABLE 6.   DRILLING ACTIVITY AND NUMBER OF PRODUCING
AND ABANDONED OIL WELLS IN THE UNITED STATES,
1951–64

Year	Total new wells drilled[a]	New oil wells completed[a]	Service wells drilled[a]	Exploratory wells drilled[b]	Number of producing oil wells at year end	Oil well abandonments[c]
1951	44,516	23,453	1,380	11,256	474,990	10,574
1953	49,279	25,762	1,262	13,313	498,940	13,278
1955	56,682	31,567	760	14,937	524,010	9,968
1957	55,024	28,012	1,409	14,707	569,273	8,651
1959	51,764	25,800	1,670	13,191	583,141	11,451
1960	46,751	21,186	2,733	11,704	591,158	15,434
1961	46,962	21,101	3,091	10,992	594,917	16,977
1962	46,179	21,249	2,400	10,785	596,385	16,224
1963	43,653	20,288	2,267	10,664	588,657	14,363
1964[d]	45,236	20,620	2,273	10,747	588,225	n.a.

[a] Total new wells includes oil plus gas plus condensate plus service wells plus all dry holes.

[b] Includes new field and new pool wildcats, new pool tests, and outposts. Exploratory wells drilled include successful and unsuccessful efforts.

[c] Number of stripper-well abandonments. May not be complete coverage for all oil wells.

[d] Preliminary.

n.a.—Comparable series not available.

SOURCES: Number of total wells, oil wells, and producing wells from U.S. Bureau of Mines and API; exploratory wells from American Association of Petroleum Geologists; abandonments from IOCC, *National Stripper Well Survey*.

might be that in most market-demand states secondary-recovery projects are exempt from market-demand restrictions or are given bonus allowables. Also, in some states pressure-maintenance projects are given bonus allowables. Thus, because of excess capacity and the resultant restrictions on prorated wells, operators have sought to expand production that is totally or partially exempt from these restrictions. In doing this, however, they have continued to add to productive capacity and have thus aggravated the excess capacity situation even more. It should be added that new secondary-recovery techniques have been widely adopted with the result that heretofore unprofitable oil has become profitable to produce.

The series on the number of producing oil wells indicates that the growth rate has slowed substantially since 1957, indicating in another way the decline in drilling activity. The trend in abandonments is somewhat erratic, but in 1953 abandonments were about 2.6 per cent of producing oil wells, while in 1963 they were about 2.4 per cent of producing wells. The low figure for abandonments in 1957 may reflect the added revenue in this year that was caused by the Suez crisis-induced price increase in crude oil. This added revenue may have prolonged the life of some wells. It is difficult to conclude from these data that the rate of abandonment is increasing significantly. This is not surprising, since the disincentives to drill are somewhat different from the incentives to abandon. In most states, no matter how much excess producing capacity exists, marginal or stripper wells are allowed to produce at or near their capacities. Movements in crude-oil prices are a much more significant source of incentives and disincentives for abandonment than are changes in excess capacity which is reflected primarily in changes in allowables for prorated wells.

A note of caution should be injected about relating wells drilled or producing wells to productive capacity. It is quite conceivable that 21,000 new oil wells in 1964 could have had as much producing capacity as 31,000 new oil wells had in 1955. If new wells in 1964 are more widely spaced than the new wells of 1955, the efficient productive capacity may be greater per new well. This is not to say that one well on 80 acres would have as high efficient producing capacity as two wells combined, drilled on the same 80 acres. However, it is likely that the efficient capacity of the one well would add more to producing capacity than either *one* of the two 40-acre wells. The fact that well spacing in new fields is moving toward wider patterns may explain part of the apparent contradiction that appears with an increase in overall producing capacity in the face of declines in the number of new oil wells drilled. Statistical testing of this hypothesis is difficult because of a lack of data on gross productive capacity each year.

The growing importance of natural gas liquids warrants some additional comment. The figures in Table 4 indicate that NGL production increased from 205 million to 419 million barrels between 1951 and 1964.[10] THE NGL contribution to the supply of all oil rose from 7.4 per cent to 10.4 per cent in this period. NGL

---

[10] U.S. Bureau of Mines figures for NGL are used in Table 4. The definitions used and the resulting figures reported by the API differ somewhat, but do not change the general conclusions.

is a substitute for crude oil in many markets and is often used as refinery feedstock in the production of motor fuels and other finished products. NGL production comes from both oil wells and gas wells, but in neither case is it restricted directly. Since both oil and gas production are curtailed in some instances, NGL production may also be curtailed. It is safe to say, however, that NGL has largely escaped regulation at either the state or federal level. There are no production restrictions similar to those imposed on crude oil. Pursuance of this topic leads directly into state regulation of gas production, which goes beyond the scope of this study. The upshot of all this is that NGL production has grown, in part at least, at the expense of crude-oil production.

There is a whole gamut of incentives and disincentives working to stimulate or retard drilling and production. We will postpone a fuller discussion of this until we have examined in some detail the workings of the state conservation regulations.[11]

THE LONG-TERM NATURE OF EXCESS OIL-PRODUCING CAPACITY

Numerous articles, books, and speeches in recent years have called attention to the fact that we are not adding to our net crude-oil reserves as fast as we used to, and that the "reserve-life-index" (the ratio of reserves to production) is falling. The federal government has noted its concern in this area also. The implications of such statements often seem to be that things are changing and that we may soon wake up with a deficiency—not only in reserves, but also in producing capacity. Conclusions of this sort with respect to reserves may be reasonable, depending on one's views on necessary inventory levels for the industry. With respect to producing capacity, however, such conclusions seem ill-founded in fact and are contrary to reasonable expectations about the performance of the industry in the foreseeable future. Herein lies a major area of confusion about supplies: the differences and interrelationships between reserves and producing capacity.

We may, in the future, find that reserves are "too low" to meet certain national security criteria. The relationship between reserves

---

[11] See Chapter 9 for a discussion of incentives and disincentives.

and producing capacity is a complex one, and the two do not necessarily move together. The U.S. Department of the Interior has
perhaps summed up the problem as well as anyone: "There is,
then, a need for the development of additional reserves and an
equal need to conduct such development in a manner which does
not contribute to already excessive capacity."[12] It seems likely that
under present conservation regulations there will be additions to
producing capacity with additions to reserves. Unfortunately, the
Department of the Interior gives no suggestions for solutions to
the problem it poses. At any rate, it seems quite likely that the
domestic oil industry will have substantial excess producing capacity
for several years to come. A most perplexing problem that is just
dawning on the industry is the fact that it may be impossible to
increase the reserves-to-production ratio in the future without generating additional *excess* capacity. This in turn creates the necessity of imposing rather stringent market-demand controls and creates or maintains downward pressure on crude-oil prices. The
industry has been slow to realize that the present control systems
have built-in incentives for operators to add to producing capacity.

If we take 1964 production of 7,614,000 barrels daily,[13] and assume
that the NPC productive-capacity figure for 1964 of 11,590,000
barrels daily is roughly correct, it would take about five years for
productive capacity to equal 1964 production if capacity fell 8 per
cent per year and no new wells were drilled. As long as new drilling
adds productive capacity as fast as it is being depleted in old wells,
we must rely upon growth in the demand for domestic crude to
absorb the excess capacity. If we project a growth rate of 2.5 per
cent a year for this demand, and assume that new drilling just
offsets losses in productive capacity from old wells, it would take
about fifteen years for demand to equal productive capacity. If
drilling in any year adds to productive capacity, net of losses due
to depletion that year, the time when demand and capacity are in
balance is pushed further into the future. One may question the
use of an 8 per cent decline rate as being too low, or the 2.5 per
cent per year increase in demand as being too low, but these figures
are, so to speak, "in the ball park." We would also agree that no

---

[12] *An Appraisal of the Petroleum Industry of the United States, op. cit.*, p. 44.
[13] Recall that crude-production figures include lease condensate and thus are
somewhat overstated.

one should maintain that the ideal state is when demand and productive capacity are equal, and that one cannot ignore the logistics of supply and demand in a country as large as the United States. Yet despite all the qualifications that can be raised about the relationships of drilling to capacity and capacity to demand, it is impossible to escape the conclusion that, barring some dramatic change in demand or drilling activity, the industry can expect to live with substantial overcapacity for several years to come.

Some light is thrown upon the nature of the problem by the projections of demand in a recent study by the Department of the Interior.[14] Upon certain limiting assumptions, production of crude oil is projected at 9,315,000 barrels per day (3.4 billion barrels per year), excluding lease condensate, for the year 1975. Comparing with the NPC estimate of crude-oil productive capacity at 11,590,000 barrels per day as of January 1, 1964, if productive capacity were to remain unchanged, there would in 1975 still be excess capacity of more than 2,000,000 barrels per day.

It may also be noted that the complete prohibition of crude-oil imports would be far from sufficient to equate demand with domestic capacity. In 1964, excess capacity stood at 4,009,000 barrels per day, and crude-oil imports were at a level of about 1,200,000 barrels daily.[15] Such a restriction on imports would, of course, help domestic producers greatly since the incremental cost per barrel of expanded domestic production would be substantially below price, but it would release less than one-third of the unused capacity reported in 1964.

Finally, it must be kept in mind that in states where significant market-demand restrictions have been in effect for several years, and where many prorated wells have been held well below their capacities, there is additional capacity potential from well workovers and reservoir stimulation. The fears of many industry representatives that NPC capacity figures are overstated do not seem justified in light of this latent capacity that could easily be made available.

Our purpose is not to make forecasts of growth in demand or of rates of decline or growth in productive capacity at different as-

---

[14] *An Appraisal of the Petroleum Industry of the United States, op. cit.*, Table 28.

[15] Crude-oil imports for the year 1964 were about 439 million barrels.

sumed levels of drilling. We merely wish to show that there is a great likelihood that the domestic producing industry must live with conditions of excess productive capacity and the stresses they create. More important for our purposes, the state regulatory agencies must face up to this situation also.

The heart of the economic difficulties that have overtaken the system of regulation lies in this excess capacity, relative to the amount of domestic crude oil demanded.[16] A skeptical reader with a long memory, looking at the preceding statement, might demur. The system, he might say, was originally introduced precisely because of a state of serious overcapacity, and its primary function has always been to cope with that situation. This is true as far as it goes, and even in the early years of proration in the 1930's, several states faced the anomaly of the rapid rise of new capacity caused by prolific drilling in new bonanza fields at the same time that they could market only a small fraction of what they were already capable of producing. The current situation differs significantly from that of the 1930's. Regulatory systems, capable of exercising control, were still in their infancy. The laws and regulations were experimental in nature and became entangled in extended litigation; the industry was often rebellious at regulation; and the technical knowledge of reservoir behavior that was important for the writing of proper regulations was still largely lacking. Much of the capacity of the 1930's existed before the states really had a chance to act effectively.

The current situation of overcapacity has developed during a period in which states had well-established regulatory systems, and in which technical knowledge was vastly improved and the industry had accepted regulation. True, we have overcapacity similar to the 1930's, but we have little of the chaos, uncertainty, and ignorance. Overcapacity is easy to explain in the 1930's, but as we have noted earlier, the current overcapacity is not nearly so easily explained. One major contributing cause seems to have been the inability of state legislatures and conservation authorities to adjust to changing conditions.

---

[16] We are not unaware of arguments for "reserve producing capacity" for national security purposes. If such arguments are valid, then the significant question is whether we are minimizing national defense costs with our present system, which generates overcapacity. To our knowledge, no one has made such cost analysis.

The previous discussion has indicated that some of the fundamental problems of a conservation regulation system are to be
found in the relations among rising capacity, the investment costs
of new capacity, and the state of market demand. The problems of
the 1930's which, if prolonged, might have undermined the system,
were banished by World War II. Demand caught up with capacity
and, for a number of years after the war, a high percentage of the
rising capacity could be produced and marketed at favorably rising
prices. We have shown that this situation induced a burst of exploratory and development activity for a decade after the war which was
to carry capacity far ahead of a demand that grew quite slowly.

The figures for 1964 in Table 2 indicate that production in that
year was only about 66 per cent of capacity. Such general figures
do not, however, show the severity of the pressure of overcapacity
upon the regulatory system. A substantial proportion of existing
capacity is exempt from proration, so that the mounting overcapacity bears down the more heavily on the prorationed segment.
The case of Texas provides a striking illustration. For example,
in November, 1963, the proration picture was as follows:

Total allowables	2,828,000 barrels per day
Exempt from proration	1,248,000 barrels per day
Subject to proration	1,580,000 barrels per day

While they have no accurate way of measuring it, the Texas authorities estimate that the wells in the prorated category had, at
this time, an efficient sustainable capacity of around 3,700,000 barrels per day. On this basis, the allocated allowables came to about
43 per cent of allocated capacity or, to put it the other way around,
capacity was 232 per cent of allocated allowables $\left(\frac{3,700,000}{1,580,000}\right)$.[17] We
do not have a similar breakdown of figures for other states, but,
presumably, a comparable situation exists in Louisiana and perhaps
Oklahoma, although not to such a severe degree.

If capacity mounts up, while demand does not, the only recourse
for regulatory authorities is to cut back the percentage of capacity

---

[17] This calculation is not the same as that used in the following chapter in
setting allowables. Current allowables are stated as a percentage of basic "schedule
allowables." This percentage in Texas during 1964 was around 26 to 28, indicating that schedule allowables are fictitiously high as compared with efficient
sustainable capacity.

that may be produced. In PAD Districts I–IV (omitting District V, the West Coast), from 1956 to 1962, production rose by just under 300,000 barrels per day, or just under 5 per cent.[18] But production actually declined in Texas, Oklahoma, Kansas, Illinois, and Colorado, out of the top ten producing states. The decline in Texas was 461,000 barrels per day, or 15 per cent. Louisiana, Wyoming, and New Mexico were on the upgrade. Louisiana production rose by 62 per cent, Wyoming by 40 per cent, and New Mexico by 24 per cent. The impact of rising capacity is very unequal as among states.

SOURCES OF ECONOMIC INEFFICIENCY

"Economic efficiency" can be concisely defined as cost outlays no larger than are necessary to achieve a given production result. By this standard economic criterion, investment outlays that are in excess of what are necessary constitute misallocation of resources, a form of "economic waste." This is clearly the state of affairs in the petroleum-producing industry. The industry is by no means unique in exhibiting such waste; there are innumerable instances throughout the economic system. But it is generally presumed that, where competitive forces are active, the tendency will be for low-cost firms to preempt the market, for prices to be regulated by costs, and for investment to be directed toward those channels where it can most effectively increase production in response to consumer demand. This presumption provides the main economic argument for a system based on free private enterprise. Among the unique things about the petroleum industry are (1) the great excess of investment in producing facilities, (2) a system of public regulation that often seems deliberately to induce such excess investment, and (3) an ambiguous but relevant national security role that justifies for some people the excess investment and that is jeopardized by excess investment in the eyes of other people.

The whole idea of wasteful investment is sometimes challenged. To quote an exasperated critic of wide well spacing: "What is all this talk about wasteful investment? It pays me to make the investment and that is what free private enterprise is all about. You

---

[18] The figures cited in this paragraph are from U.S. Department of Justice, *Report of the Attorney General Pursuant to Section 2 of the Joint Resolution of August 7, 1959, Consenting to an Interstate Compact to Conserve Oil and Gas* (Washington, D.C., 1963), p. 22. (Based on U.S. Bureau of Mines data.)

never hear anyone complaining about all the unnecessary filling stations." This attitude defines an issue. There is certainly economic waste from society's viewpoint in unnecessary filling stations and elsewhere in competitive industries. It would be impossible to prevent this without severely regimenting market structures and restricting free entry to economic occupations. What differentiates the petroleum industry is that it is already regimented under public authority, and that regulatory agencies implementing conservation statutes in effect *require* operators to make wasteful investments, as a necessary condition for capturing a share of the underlying oil and for sharing in the total market of the state, no matter how much they might wish to avoid the necessity. At the same time, by restricting production to achieve market stability, the public authorities do in effect, in a collective sense, support prices that make the investment profitable.[19] It should be said that many operators appear not to want to be relieved of the necessity of wasteful investment, but that does not change the facts.

The domestic oil producing industry in the market demand states is one in which a rather large number of firms are active participants, but in which these firms are forced, by regulation, to cooperate in exercising control over supply and consequently over price. In such a situation, relatively high-cost firms can continue to exist under the price "umbrella," while low-cost firms reap substantially higher profits. If, at the same time that supply is being limited and prices held up, entry by new firms into the industry is possible, then the entry of those firms will lead to excess capacity in the industry and to some loss of output and earnings by those already in it. While additional investment adds up to an accumulation of idle capacity, this need not prevent the new investment from being remunerative to the owners as long as they are provided with production quotas that will provide revenues large enough to more than cover their costs. In the same way, older high-cost firms are not driven from the industry as long as the revenues from their quotas exceed their mere operating and maintenance costs. A "nor-

---

[19] The view that overcapacity is, in fact, an equilibrium situation because of state regulations and federal import controls is systematically developed in A. E. Kahn, "The Depletion Allowance in the Context of Cartelization," *American Economic Review*, Vol. 54 (June, 1964), pp. 286–314. See Chapter 8 below for a discussion of prices, the market mechanism, and the role of regulatory commissions in the market.

mal" return on their past "sunk" investment is not required to keep them alive and it may, indeed, have long since been recovered.

Thus we have an industry in which supply is not controlled by voluntary cooperative action of members of the industry, but it is effectively controlled collectively, though not cooperatively, by regulatory agencies through proration.[20] Given this supply, a level of market prices is maintained and, given these prices, new capacity will be provided if permitted and if the assigned quotas are sufficient to cover the capital and operating costs of finding and developing additional oil. Though arising out of private investment decisions, the amount of excess capacity is in an indirect way determined by the provisions of the state conservation statutes and by the regulatory authorities through the incentives provided by their control over pressure maintenance and secondary recovery projects, well spacing, and assignment of production allowables. The private calculation with respect to drilling development wells is whether the prospective quota is sufficient to justify the drilling investment, in terms of risk, annual revenue, and length of payout period.[21] If capacity increases while demand fails to rise correspondingly, the necessary adjustment is a curtailment of the permitted rate of output of both new and existing prorated wells, which raises the unit cost of output and extends the payout period.[22]

The tendency in such a system, if demand does not rise to validate larger productive capacity, is in the direction of a stalemate. Higher unit costs and lower anticipated profits on investment in new capacity will induce declining development outlays; this in turn, unless offset by some special stimulative measures, will also carry down exploratory investment. A firm rise in demand, permitting larger quotas, can only salvage the profit prospects temporarily. Higher profits would again stimulate drilling and the consequent growth of capacity. The only approach to keeping new investment profitable is through lower costs—an approach that would lead in

[20] It should be noted that the part of supply from foreign sources is also controlled by means of the oil-import-control program. This discussion ignores imports for the time being.

[21] A more technical statement of this relation in the language of economic theory would be that, as long as price is held above the short-run and long-run marginal costs of efficient producers and entry is permitted, investment in additional capacity will be induced.

[22] As was noted earlier, the cutback may not hit certain classes of wells exempt from production restrictions.

the direction of greater economic efficiency through reducing the amount of investment in idle capacity. Another approach suggested is to attempt to get prices up, but, if successful, this leads to greater capacity also. Certainly the industry has, through innovation, reduced costs. Although some progress has been made, it has been much less successful in promoting a major overhaul of regulations with an eye toward cost reduction.

The assumption in the preceding description that capacity increases more rapidly than slowly rising demand has been approximately true of crude oil for a number of years past for the country as a whole. It is not precisely true from state to state since, where market demand is available, some states have been able to increase their shares at the expense of other states and especially at the expense of Texas. Put more precisely, purchasers of crude have changed their purchases among states. This, of course, changes the drilling incentives as between one state or another. The size of the drilling units permitted or prescribed and the formulas for allocating production to new reservoirs and wells may also provide a differential stimulus or deterrent to exploratory or developmental drilling from one state to another. These differences among states are of some importance in relation to the competitive interest of states, but they affect in minor degree the essential characteristics of the general situation. Since some secular increase in the demand for crude oil may be expected to occur, not wholly offset by increased imports, the assumption of a relatively slow rise of production need not be permanently applicable. Insofar as production does increase from existing capacity, it will naturally benefit the producers and may delay the movement of the system toward stalemate. However, as present production becomes more profitable, this in turn creates the incentives to drill unnecessary additional wells, which in turn creates more excess capacity. Thus, under current regulations, increase in production will ultimately stimulate an increase in capacity.

Some members of the industry entertain the hope that the situation described is temporary and will in due course remedy itself without the necessity for any marked change in regulatory practices. This hope is based on various statistical projections of demand which suggest that, in spite of the inroads of imports and natural gas, the demand for crude oil may move strongly upward in the not too distant future. However, as we noted on page 112, Depart-

ment of the Interior projections of demand suggest a heavy inci-
dence of overcapacity at least for the next decade. Additionally,
we have noted that incentives may be created that will induce
additions to capacity. In any case, operators are feeling the cost
pinch, and regulatory agencies are engulfed by the dissatisfaction
of operators over relatively small allowables. As one regulator sadly
says, "When everyone produces at near capacity, no one complains
very much. When the market is slack, everyone blames the other
fellow." Since the domestic producing industry appears destined to
have considerable excess producing capacity for several years to
come, the industry as well as many regulatory authorities are calling
for adjustments in conservation regulations to fit the situation.

Even if the regulator can see his problem largely in terms of
keeping a dissatisfied clientele off his neck, the basic problem is a
more fundamental one. It is one of keeping the industry dynami-
cally engaged in the discovery of new reserves while at the same
time halting and reversing the mounting load of excess capacity.
This calls for progress in the direction of increasing economic
efficiency through lower costs.

It is generally true that any addition of reserves brings new productive capacity into being. The important fact is how much producing capacity is generated when new reserves are added. At the
end of 1951, U.S. proved crude-oil reserves stood at 25,268 million
barrels while productive capacity was 6,727,000 barrels daily. Thus,
we had 3,756 barrels of reserves per barrel of daily producing capacity. By the end of 1963, proved crude reserves stood at 30,970
million barrels and productive capacity at 11,590,000 barrels daily
for a ratio of 2,672 barrels of reserves per barrel of producing capacity. Until some means are found to reverse this trend and have
reserves grow more rapidly than capacity, no amount of exhortation
is likely to stimulate the accelerated exploration for the new reserves that is being called for by government and industry spokes-
men alike. Certainly, in this sense, some regulatory practices are a
threat to and not a bulwark of national security. There are hopeful signs in some recent changes in regulations, such as the new
yardsticks in Texas and Louisiana, wider well spacing, and more
pooling. In our opinion, they do not go far enough.

The excessive investment in producing capacity has two aspects:
(1) the drilling of unnecessary development wells and (2) the restricted use of developed capacity. If one well would efficiently drain

160 acres, 40-acre spacing multiplies the drilling cost by four. If production is prorated to 50 per cent, the drilling cost per unit of output is so much further inflated. Moreover, the amount actually produced might have been efficiently produced on much wider spacing than 160 acres. As everyone knows, great investment-cost savings could be accomplished through wide spacing in many situations.[23] The practical questions are (1) whether the cost-price squeeze will become sufficiently severe to overcome the objections of those who have heretofore favored closer spacing, and (2) whether the proration formulas will be changed to provide an overwhelming interest of the operators in larger drilling and producing units.

The economic logic of the situation is not met simply by eliminating small-tract drilling and enforcing some conventional unit like 40 acres. If wells drilled on 40-acre tracts are immediately prorated back to one-half or one-third of efficient capacity, the situation is still one of excessive investment relative to output. The only solution corresponding to economic criteria is one in which wells are so spaced that they can be permitted to operate close to 100 per cent of efficient capacity for the pool. It is more efficient to have one well on 80 acres producing at its MER than to have two, four, or eight wells producing at 50 per cent or less of efficient capacity.[24] Even production at MER on wide spacing does not reach the ultimate in efficiency if there remain unexploited opportunities to increase recovery through unitized pressure maintenance programs. While it would no doubt be foolish to expect anything closely approximating an ideal of efficient development, it is necessary to recognize that the obstacles consist mainly of conventional attitudes and prevailing practices that are capable of being modified in the interest of cost reduction. The practical question is how far the pressure of avoidable costs will override the obstacles, and how rapidly cost-reducing changes will be made.

---

[23] This discussion assumes stable future prices and ignores the cost of waiting, that is a discount rate, if production is spread over a longer period with fewer wells in a reservoir. It is clear that an individual operator contemplating the optimum number of wells (his investment) must take into account prices, a discount rate, and expected allowables, among other things. These considerations, however, do not alter the main thrust of the argument presented here, which is from an aggregate industry or social view.

[24] The MER is a concept applicable to a pool and not to the individual wells within a pool. There is, however, for a pool, an efficient producing rate that can be divided among the wells in the pool.

There are, it must be recognized, insuperable obstacles to placing the whole industry on an economically efficient basis at an early date. Even if the most perfect canons of efficiency were applied to the development of new reservoirs, nothing much could be done about the unnecessary investment costs sunk in reservoirs already more or less fully developed. It is, of course, in these reservoirs that all of the present excess capacity exists. However, it would be possible in many cases to plug unnecessary wells to save operating expenses and maintenance costs.[25] These savings could be substantial in the aggregate. Such proposals have been made from time to time. In 1962, two large operators in the East Texas Field proposed that about seven-eighths of the wells in that field be abandoned and the field be produced at the same rate from the remaining wells. The Texas Railroad Commission turned down the proposal on the grounds that it would depress the local economy that was dependent on payrolls in the field.[26] It must be obvious how strongly vested local interests stand in the way as political barriers to improvement in the operating efficiency of already developed fields. A less ambitious and less efficient proposal was made in 1965 by another large operator and supported by most of the other major producers in the field. This proposal could conceivably reduce the number of wells from 17,200 to 9,500 and increase the present average spacing of 4 to 5 acres per well to 15 acres per well. One witness was reported as testifying that "there are at least four to eight times too many wells, and it may be as high as 16 times."[27] Such statements have been made before. One expert has stated that only 1,500 wells are needed to drain the field efficiently.[28] At the time of this writing the commission denied the request made in this proposal.

One past result of proration is to give existing wells what amounts to a vested interest in their present quotas, subject to general changes in the level of allocated production. As long as these vested interests are respected, and with due regard to natural decline rates and increase of demand, the only way in which overcapacity could

[25] The investment costs in pumping equipment put on wells as the natural flow diminishes would also be saved since such equipment is normally installed on each producing well in the field.

[26] *Oil and Gas Journal*, August 22, September 26, and December 5, 1962.

[27] *Oil and Gas Journal*, June 21, 1965, p. 100.

[28] Thomas C. Frick, *Journal of Petroleum Technology*, Vol. 10 (1958), p. 12.

be reduced would be to limit the growth of capacity in new reservoirs. We must first examine the regulatory systems in more detail before facing up to possible solutions to the three-pronged problem of how, simultaneously, to (1) introduce efficient low-cost development of new reservoirs, (2) reduce excess capacity, and (3) retain strong inducements to exploration.

THE RELATIONSHIP OF MAJORS AND INDEPENDENTS

Affecting the somewhat discordant relationships of nonintegrated "independent" producing companies to the integrated "major" companies, some problem areas arise between these two groups because of the existence of overcapacity in the industry. Virtually all major companies are partially dependent on purchased crude oil to meet their refining needs. This is the primary market for crude oil produced by independents. Overcapacity in production means that all prorated wells, whether owned by majors or independents, are restricted in a given state. Restriction, as was indicated above, raises costs per barrel of production. A major company, in filling its crude-oil needs for refining, would like to take oil from its own wells, as long as the incremental cost of this production is less than the price. In situations of production well below capacity, incremental costs are invariably substantially below price; thus the incentives for majors to use their own production are correspondingly strong.[29] Given the system of proration plus ratable take laws which require that purchasers buy ratably in a pool among the wells to which they are connected, it is impossible for a major to utilize all of its own producing capacity.[30] It can, however, do several things that will move in that direction. It can shift exploratory and development activity to areas in which the major owns a larger part of the producing acreage. This may be in the same state or in another state geographically. This will enable the major to satisfy its crude oil needs to a larger extent from its own wells and less from wells of independents and other majors.

---

[29] There is the additional incentive of percentage depletion allowed for federal income tax purposes on income from production. See Chapter 8 for a discussion of this.

[30] We should again note that ratable take laws are not often used. The threat of possible use is, however, always present.

The major may also dispose of pipeline gathering facilities in areas in which it owns little production. This allows the major to concentrate its buying in areas where it owns more allowed production. If an independent gatherer purchases a major's gathering system in a particular area but still ties into the major's trunk pipeline system, the major can buy or not buy from the gatherer as its needs dictate. Ratable take laws do not apply to sales from a gatherer to a trunk pipeline.

If two adjacent states have substantially different regulations vis-à-vis ratable take or interpool allowables, it may induce a major to shift to the state that permits more favorable treatment of the major's own production. This clearly is a long-run phenomenon but, since overcapacity must also be viewed as a long-term situation, it may be quite relevant. It is good economic sense on the part of a major to try to minimize his crude-producing costs. It is also quite logical for independents to react adversely to a situation in which their markets are threatened or in which price softness is increasing. Overcapacity helps to explain some of these internal industry strains.

IMPORT CONTROLS AND OVERCAPACITY

In arguments for more severe restriction of imports, the rising level of imports has at times been advanced as an important cause of excess capacity. While it is true, of course, that lower imports would provide a larger market for domestic production, no important relation between imports and the rise of excess capacity can be established.

Using the data for Districts I–IV where practically all the excess capacity is located, from 1951 to 1963 imports of crude oil and refined products (excluding residual fuel oil which has little relation to the demand for domestic oil) rose from 508,000 to 996,000 barrels per day, or an increase of 488,000 barrels per day.[31] From 1951 to 1964, as shown in Table 2 above, excess capacity increased by more than 3,100,000 barrels per day. While the volume of imports into the eastern United States cannot be entirely disassociated

---

[31] *An Appraisal of the Petroleum Industry in the United States, op. cit.,* Table 21.

from the play of other forces upon producers in the main producing areas, it may largely be discounted in relation to the rise of producing capacity.

A further word needs to be said, however, about the relation of import policy to the long-run problem of excess capacity. Since 1959, import quotas have been set by the federal government, the present rules fixing the total volume for Districts I–IV at 12.2 per cent of domestic production of crude oil and natural gas liquids. The program was initiated under the rubric of national security as an exception to general U.S. trade policy favoring freer international trading under the rules of the General Agreement on Tariffs and Trade. The stated purpose was to add to the "health" and "vigor" of the domestic industry with a view to increasing the incentives for the search for new reserves, designed to keep the United States in a relatively more self-sufficient position for meeting national emergencies.

From what has been said earlier, the country obviously has the producing capacity to meet any likely emergency requirements in the early future. The end in view must therefore be thought of as provision for the possible needs of a relatively more distant future. To the extent, however, that import restriction is effective in its stated purpose, it will stimulate exploratory effort in the present. And, since addition to reserves is normally accompanied by additions to producing capacity, it will help to perpetuate the current state of excess capacity.

There is a paradox here to which no clear answer is visible. State regulatory practices, as we have seen, inherently have had the effect of stimulating overcapacity. Federal policy is designed to stimulate the search for reserves, which would have the normal effect of exaggerating that tendency. There is not in sight any change in regulatory practice that would have the effect of preventing an accelerated addition to reserves from having the correlative effect of enlarging capacity. How important the question might become depends upon the extent to which federal policy does, or might, stimulate the discovery of new reserves. For the present the attitude of the federal authorities appears to be one of sitting tight on the existing 12.2 per cent restrictions formula, waiting to see what consequences may arise from it. Meantime, domestic producers are limited in their ability to argue plausibly for more severe restriction by reason of the large existing overcapacity. If they plead for lower imports as a means of relieving the impact of excess capacity, the answer must be that very slight relief is available in that quarter and that the disease must be attacked at the source.

The subject of this chapter—excess producing capacity—is not so much a separate topic as a central point at which a number of features of the regulatory process may be seen converging. It arises out of the whole range of incentives that are presented for capital investment. These flow, on the one hand, from the rules governing drilling rights and reservoir development that we have reviewed in earlier chapters and, on the other hand, from the rules governing production allowables under the proration system to which we turn our attention in the next chapter. Supporting the whole structure of incentives is the stable structure of high prices created by control over production.

Barring a growth in demand sufficient to validate the excessive capital investment, the tendency of the system is toward a declining rate of use of existing and new capacity and consequently higher unit costs and a declining rate of return on new investment. The possibility that rising demand will keep the system intact depends upon the extent to which alternative sources of energy are able to invade its markets. Since the alternative sources are crowding in at the competitive margins, the avenue of escape may have to be through measures that would reduce costs—of which the most obvious means would be the cessation of new investment in unnecessary wells and unused capacity. But to achieve this result would require a drastic overhaul of the regulatory practices that have led to the present situation.

Chapter 6

The Proration System

Proration consists of the rules and procedures by which a regulatory agency determines the total crude-oil production for a state and allocates the total among the various reservoirs and to the producers in each reservoir.[1] Exempt and special categories of producing wells and/or fields must be defined and, for the remaining wells or fields, formulas must be devised by which, from a base allocation figure, their shares in the total can be calculated by applying a percentage. Rules must be devised that attempt to assure that the quotas assigned to individual wells will find a purchaser. Once the governing rules are in effect, the operation of the system becomes largely a matter of administrative routine.

Another important aspect of the proration process is the way in which it reacts upon the rise of new productive capacity. The formulas governing allocation play an important role in defining the strength of the incentives to drill discovery wells and to the development of producing capacity in both new and old fields. The proration rules must not, therefore, be regarded as simply a way of dividing up a restricted total amount of production among producers, but as a part of the mechanism through which the whole process of discovery and development is influenced.

---

[1] As noted in footnote 1, Chapter 3, legally allowables are assigned to producers and not to tracts or wells. As we shall show later, the number of wells often determines the allowable for a field. For our discussion we will refer to allowables assigned to wells, with the understanding that it is legally the producer who receives the allowable.

There are two distinct reasons why proration may be initiated. One is that, even though there is an available market for larger output, the production from reservoirs needs to be restricted in order to safeguard the driving mechanisms and insure large ultimate recovery. This may be called MER proration, meaning that production is restricted according to some calculated "maximum efficient rate" dictated primarily by specific reservoir characteristics.[2] If universally applied to all reservoirs in a state, the statewide total would be a purely derivative figure, made up of the controlled outputs of the separate reservoirs.

The other type of proration, commonly called "market-demand proration," consists of procedures by which the state regulatory agency restricts statewide production in line with an estimated demand in the next ensuing period—this state total being allocated back to reservoirs and individual wells. This is the type of proration that is of special importance under existing circumstances of excess producing capacity, and the present section will be devoted mainly to describing the mechanics of some state systems now in operation.[3] As we saw earlier, the historical occasion for introducing proration was the development of great excess producing capacity; this same situation today gives it continuing importance.[4]

It is not to be inferred that the two types of proration—MER and market demand—have any necessary connection with each other. It would be possible to interrelate them if each reservoir were given an "MER rating," and cutbacks to market demand were proportioned to these ratings. This is not what happens, except in special situations. The MER concept could, nevertheless, play some part in two different connections: (1) In market-demand states, if application of the proration formula permitted production beyond an efficient rate, the formula could be changed in application to particular reservoirs to prevent this. We are not aware that any state systematically attempts this procedure. (2) In states that do not practice market-demand proration, freedom to produce could be

---

[2] The MER concept is discussed below in Chapter 7.

[3] California, which has self-regulation by industry, uses a form of MER prorationing. This system will not be discussed since, as a rule, overcapacity is not a problem there, and we are primarily concerned in this paper with public regulation.

[4] As we noted in Chapter 4, excess production as well as excess producing capacity characterized the 1930's. Today we have excess capacity only.

limited by application of the MER principle to individual reservoirs. More will be said about this at a later point.

A general discussion of proration is somewhat handicapped by the diversity of regulations. In 1964, there were thirty-two oil-producing states; twelve of these have "market-demand" statutes that allow or require the regulatory body to limit production to reasonable market demand. Five of the seven largest producing states are included among these twelve. These are, with their respective ranks by 1965 production, Texas (1), Louisiana (2), Oklahoma (4), New Mexico (6), and Kansas (7). California (3) has a system that is largely self-imposed and self-administered by the industry, and Wyoming (5) has a state-administered system that excludes market-demand proration. The top five market-demand states have about 70 per cent of U.S. production and the seven smaller market-demand states[5] add another 4 or 5 per cent. Thus we find market-demand proration geographically concentrated in the Southwest, but covering a large major fraction of national oil production.

Among states following a market-demand plan, the formulas used are very different. And even in some states not using such a plan, analogous problems of control arise. It is therefore difficult to generalize about proration in the nation as a whole. Despite the handicap of diversity, we will attempt general description with diversified illustrations mainly from three states that have market-demand proration—Texas, Louisiana, and Oklahoma.

PRORATION AND PREVENTION OF WASTE

Market-demand proration takes places under conservation statutes that are universally couched in terms of "preventing waste." The definition of what constitutes waste plays a part in evaluating proration. Each state employing a proration system has statutes, rules, and regulations setting forth in considerable detail the framework and working rules of the system. Each state also has a regulatory authority to which the legislature has delegated powers to administer and police the day-to-day operations of the system. The relevant state laws as well as the rules and regulations of the conservation commission are subject to judicial review in the state

---

[5] Alabama, Florida, Iowa, Michigan, North Carolina, North Dakota, and Washington.

courts and, in limited areas, in the federal courts. Thus the courts have played an important role in shaping the proration systems in the individual states. While court interpretation is of legal issues, the impact has often been economic.

A discussion of the concept of waste involves the entire conservation concept, but it is helpful in providing a springboard for the distinction between (1) systems that are nominally concerned only with problems of so-called physical waste and (2) the systems that have a wider scope and consider economic waste as well. The distinction between physical and economic waste is not satisfactory, since all waste, as the term is used in the oil industry, is economic waste. There are, however, situations in which the alleged waste does not involve the loss of physical barrels of oil above ground or underground, or the loss of reservoir energies, but involves endangering market stability and the consequent economic losses caused by this instability. This distinction will become clearer as the discussion proceeds. The Texas statute states that the term "waste" among other things shall specifically include:

(a) The operation of any oil well or wells with an inefficient gas-oil ratio . . .

(b) The drowning with water of any stratum or part thereof capable of producing oil or gas, or both oil and gas, in paying quantities.

(c) Underground waste or loss however caused and whether or not defined in other sub-divisions hereof.

(d) Permitting any natural gas well to burn wastefully.

(e) The creation of unnecessary fire hazards.

(f) Physical waste or loss incident to, or resulting from, so drilling, equipping, locating, spacing or operating well or wells as to reduce or tend to reduce the total ultimate recovery of crude petroleum oil or natural gas from any pool.

(g) Waste or loss incident to, or resulting from, the unnecessary, inefficient, excessive or improper use of the reservoir energy, including the gas energy or water drive, in any well or pool . . .

(h) Surface waste or surface loss, including the storage either permanent or temporary of crude petroleum oil or the placing any product thereof, in open pits or earthen storage, and all other forms of surface waste or surface loss, including unnecessary or excessive surface losses, or destruction without beneficial use, either of crude petroleum oil or of natural gas.

(i) The escape into the open air, from a well producing both oil and gas, of natural gas in excess of the amount which is necessary in the efficient drilling or operation of the well.

(j) *The production of crude petroleum oil in excess of transportation or market facilities or reasonable market demand.* The Commission may determine when such excess production exists or is imminent and ascertain the reasonable market demand.[6] [Emphasis added.]

The Oklahoma statute is less detailed, but is as all-inclusive. It states that:

The term "waste," as applied to the production of oil, in addition to its ordinary meaning, shall include economic waste, underground waste, including water encroachment in the oil or gas bearing strata; the use of reservoir energy for oil producing purposes by means of methods that unreasonably interfere with obtaining from the common source of supply the largest ultimate recovery of oil; surface waste and *waste incident to the production of oil in excess of transportation or marketing facilities or reasonable market demand.*[7] [Emphasis added.]

The Louisiana statute repeats much of the same language:

"Waste," in addition to its ordinary meaning, means "physical waste" as that term is generally understood in the oil and gas industry. It includes: (a) the inefficient, excessive, or improper use or dissipation of reservoir energy; and the location, spacing, drilling, equipping, operating, or producing of an oil or gas well in a manner which results, or tends to result, in reducing the quantity of oil or gas ultimately recoverable from a pool; and

(b) the inefficient storing of oil; *the producing of oil or gas from a pool in excess of transportation or marketing facilities or of reasonable market demand;* and the locating, spacing, drilling, equipping, operating, or producing of an oil or gas well in a manner causing, or tending to cause, unnecessary or excessive surface loss or destruction of oil or gas.[8] [Emphasis added.]

From the language of the statutes, it is clear that each is a market-demand state. Production in excess of transportation, marketing facilities, or reasonable market demand is prohibited.

Wyoming, a state that does not have a market-demand provision, has the following statutory language:

The term "waste," in addition to its ordinary meaning, shall include:

The term "waste" as applied to oil shall include underground waste, inefficient, excessive or improper use or dissipation of reservoir energy, including gas energy and water drive, surface waste, open pit storage and waste incident to the production of oil in excess of the producer's above-ground storage facilities and lease and contractual requirements, but excluding storage (other than open pit storage) reasonably necessary for

---

6 Title 102, Revised Civil Statutes of Texas, Article 6014.
7 52 O. S. A., Sec. 86.2.
8 Louisiana R. S. 30:1.

building up or maintaining crude stocks and products thereof for consumption, use and sale.[9]

While this statute prohibits waste, it does not specifically by words include as waste production in excess of reasonable market demand. It is perhaps possible to interpret the phrase "waste incident to the production of oil in excess of the producer's above-ground storage facilities and lease and contractual requirements" as production in excess of market demand. The Wyoming statute does have virtually the same prohibitions on "physical waste" as the market-demand states.

With respect to implementing the prohibitions against waste, most market-demand states have a statutory provision similar to that in Texas:

"The Commission may consider any or all of the above definitions [of waste], whenever the facts, circumstances or conditions make them applicable, in making rules, regulations or orders to prevent waste of oil or gas."[10]

And specifically, "The Commission may determine when such excess production exists or is imminent and ascertain the reasonable market demand."[11]

Finally, the Commission is charged with the task of making waste prevention equitable. With regard to proration or production restriction, the statutes will often contain a provision similar to the one found in Louisiana:

The commissioner shall prorate the allowable production among the producers in the pool on a reasonable basis so as to prevent or minimize avoidable drainage from each developed area which is not equalized by counter drainage, and so that each producer will have the opportunity to produce or receive his just and equitable share subject to the reasonable necessities for the prevention of waste.[12]

PHYSICAL WASTE AND ECONOMIC WASTE

The statutory provisions on waste quoted above show how tightly regulation has been tied to a concept of "physical waste." "Eco-

---

[9] Session Laws of Wyoming, 1951, Chapter 94, Sec. 13a.

[10] Title 102, Revised Civil Statutes of Texas, Article 6014.

[11] *Ibid.*

[12] Louisiana R. S. 30:11 (B). It should be noted that so-called "ratable-take laws" are somewhat different from the above quotation. Ratable take places a burden on purchasers to take ratably among wells in a field. See Chapter 7.

nomic waste" is seldom mentioned and, when mentioned, is not defined. This fact emphasizes the confusion to which we pointed in Chapter 1 about the concept of "conservation." The regulatory authorities have no working concept of "waste," so that it becomes whatever they find themselves engaged in "preventing." As against this inchoate situation, the economists have attempted to provide a theory in which waste is defined in relation to a test of economic efficiency.

The type of industry thinking on the subject that is so perplexing to economists may be illustrated by a quotation from the 1964 IOCC study:

The conservation program, especially limitation to market demand, has been criticized on the ground that its primary purpose is to prevent economic waste, not physical waste. The meaning of "economic waste" is not clear. "Economic waste" may be defined for this discussion as the loss of, the destruction of, or the failure to use, oil or gas that has value, provided that the cost of saving the irreplaceable resources of oil or gas is economically justified.

Operations that reduce ultimate recovery that could be obtained economically, or operations that otherwise cause physical waste of oil and gas, as by storing oil in open pits, or letting gas escape into the air, are examples of physical waste. They also are examples of economic waste.

The point already has been made that limitation of production of oil to market demand prevents physical waste. It also necessarily prevents economic waste as an incident to the prevention of physical waste; but, of course, does not prevent all economic waste.

The Oklahoma and Kansas statutes are the only ones that include "economic waste" in the definition of waste that the administrative agency is authorized to prevent, but there is no definition of the term, and the administrative agency has not undertaken a program of preventing "economic waste" except as an incident to prevention of physical waste.

If there is no control over production; if production is excessive because operators produce to capacity or as they please; and if stocks exceed the level necessary to carry on business, as was typical of many periods in the past, then prices are unstable and tend to fluctuate widely. However, if regulation or lack of regulation results in a lower or higher price level, it does not follow that economic waste occurs. If regulation prevents physical waste, as by restricting production to MER's, and incidentally a higher price level occurs, it does not follow that the purpose or effect is to prevent economic waste. The subject is discussed in the next subsection.

If there is overproduction, but proper regulation under conservation statutes takes place, including limitation of production, when necessary, so that it will not exceed the demand, the program is designed to prevent physical waste, as has been pointed out. The purpose of the regulation is to prevent physical waste in a reasonable, effective way, and any effect on prices and economic waste is incidental. Courts have so held, and the principle is now considered to be well established.[13]

We hesitate to view a passage such as this as a definitive statement by the industry on the definitions of and distinctions between physical and economic waste. However, it very likely represents a broad consensus of industry thinking. It can be interpreted as being virtually the same as the economist's definition if we add a time dimension. It speaks in terms of waste resulting when values are lost which on a cost basis could be economically justified or saved. It is also pointed out that physical waste is economic waste—a point upon which all economists would agree. Economists, in fact, would insist that physical waste is waste only in an economic sense and in no other sense. The remainder of the quotation is extremely confusing and, in fact, appears to us to be contradictory to the definition of economic waste given and the equating of physical and economic waste.

Physical waste as the term is used by the industry can have no other meaning than economic waste. There has yet to be a reservoir from which all the oil has been recovered. Is this oil remaining physical waste? The industry would say no, that all that is required is that "avoidable" physical waste be eliminated. But as soon as this is said, the industry is thinking in economic terms. Much more oil could be recovered if we disregard the cost of recovery. The industry must reply that oil that costs more to recover than it returns to the producer in revenue is not wasted if left in the ground. By saying this the industry has put waste in purely economic terms: cost and price.

Within the regulatory process, however, the economic principle is never pushed to its logical limit. The economic principle, for example, would prescribe that waste occurs if costs can be reduced without changing the benefits, or if price could be reduced without lowering the available supply. This principle cuts very close to the "stripper well" controversy where, in support of protecting high cost sources, the appeal is made to a purely physical concept of not wasting "a precious natural resource," without regard to the avail-

---

[13] IOCC Governors' Special Study Committee, *A Study of Conservation of Oil and Gas in the United States* (Oklahoma City, 1964), pp. 91–92.

ability of cheaper oil. The difficulty with the industry usage of the waste concept is that its economic component is undefined, and therefore provides no test for the dividing line between wasteful and unwasteful, or efficient and inefficient.

The great advantage of the economists' concept is that it provides a precise way of thinking about efficiency, placing waste and inefficiency in the same context. This does not mean that regulatory practices need always be directed to the goal of efficiency. There may be other motives or goals: national security at any cost, recovery of the maximum amount of oil from domestic deposits at any cost, the maintenance of inefficient units in the industry at any cost. If these or some other motives are appropriate, it is essential that their departure from economic cost principles be made explicit. The alternative costs of noneconomic uses of resources are part of the basis for deciding what the most desired uses are.

It is hard to understand the obsession of the industry and its regulators to cling to the phrase physical waste. It beclouds the relevant issues of conservation and results in interminable argument over what is and what is not physical waste or economic waste, or both. Possibly one reason for keeping it in the central place is a purely traditional one: oil men grow up with it. But beyond this there is a real reluctance to adopt a clear concept of economic waste. This may stem from the fact that much of what regulatory statutes say and regulatory bodies do does not correspond to such a concept. As we have seen, in many situations they require wasteful investment in excess capacity, and they often inhibit methods of reservoir development that would maximize ultimate economic recovery. They are therefore likely to think of the economic waste concept as an instrument with which critics may attack the system, rather than as an analytical device. It may well be that the smoke screen produced by the industry by its devotion to the concept of physical waste has done more to arouse suspicion and criticism than would an informed and enlightened analysis by the industry of how precisely regulation impinges on prices and costs and what the implications are for conservation. We suspect also that this type of open discussion by the industry and its regulators would bring to light ways in which regulation could bring relief to some of the economic ills that currently afflict the industry. If, however, regulators openly recognized the relation between waste and economic efficiency, they would be committed to a large degree of self-criticism.

DETERMINATION OF THE STATE ALLOWABLE

Given the definitions of waste and the mandate to prevent unnecessary waste, a state commission must then establish through rules and regulations the tools with which to work and a system in which to use the tools. While there are some statutory guides to help in forging the tools and establishing the system, in general, this is left to the commissions. It is their task to put flesh on the skeleton and provide the working details of regulation.

A market-demand statute in effect makes it mandatory for a commission to have some method by which to make short-run forecasts of consumption of crude oil originating in that state. This machinery is roughly similar in the major market-demand states, and the discussion that follows is generally applicable to all these states. Once a state allowable has been established, it is then necessary to determine the share of each pool in the state total and to determine the share of each well in each pool.[14]

To determine future demand (consumption), each state commission holds periodically (monthly or bimonthly, depending on the state) a public hearing at which it hears evidence from all interested parties, including its own staff, on what consumption can be expected. The Oklahoma Rules on this point state that:

The Commission shall, on due notice and hearing find and determine the reasonable market demand for oil, gas and other hydro-carbon products produced in Oklahoma, for consumption in and outside the State for the ensuing proration period that can be produced from each common source of supply on a statewide basis without avoidable waste with equitable participation in production and markets by all operators and other interested parties.[15]

The evidence provided includes the amounts of oil that can be expected to be purchased, the amounts of crude oil and products that are currently in storage, the amounts of crude oil and products needed in storage, pipeline throughput capacities, production and storage in other states, imports, and general estimates of consumption.

---

[14] Strictly speaking, the state total is divided up among wells and not pools. In some cases, for example in unitized fields, pool rates rather than well allowables are established. However, pool allowables are determined by summing the well allowables for most fields. This is discussed below.

[15] Oklahoma Corporation Commission, General Rules and Regulations Governing Oil and Gas, Rule 701, Sec. a.

The staff of a commission puts in evidence the "nominations" of crude oil purchasers. These nominations are purchasers' estimates of what they will buy in the state during the proration period under consideration. Nominations are required of purchasers and are often made on a field basis. While there is no statutory requirement that a purchaser take all the crude oil nominated, there is considerable pressure for each purchaser to take at least what he nominates. At the hearing a purchaser may be asked to appear in person to substantiate and explain his nominations.

Since demand can be met either from current production or above-ground stocks of crude oil or products, the commission is concerned with the level of stocks and their location, i.e., whether they are in lease tanks, pipeline tank farms, refineries, terminals, etc. Changes in the levels of stocks are one of the first indications of imbalance between supply and demand and such changes are closely watched by the regulatory authorities. Stock figures reported weekly by the API[16] keep this vital information current. Purchasers of crude oil also on occasion are asked to submit information on stocks.

An outside estimate of future consumption of crude oil for the next period is available to each state from the U.S. Bureau of Mines. The estimating procedure is discussed more fully in a later section, but in general it is built up from estimates of product consumption and is quite comprehensive and accurate. This gives a commission another opinion as to near-future crude-oil needs. Some states seem to rely more heavily than others on the estimates of the Bureau of Mines, or at least give them more recognition in hearings and orders. Oklahoma, for example, always notes this Bureau of Mines forecast and comments if its own estimate differs substantially. Texas, on the other hand, gives it no recognition in its printed monthly proration order. Every market-demand state makes use of it to some extent, even if it is no more than a checkpoint.

Table 7 shows nominations, allowables, actual production, and Bureau of Mines consumption estimates for crude oil in Texas during 1961 and 1962 by months. Allowables are set in the middle of one month for the following month, and some attempt is made

---

[16] Prior to January, 1964, the U.S. Bureau of Mines reported these weekly stocks.

to reflect minor changes in productive capacity and production that might arise because of abandonments and new completions during the month. Allowables in Texas are consistently about 12 per cent above what the commission feels should be produced.

TABLE 7.　CRUDE-OIL PRODUCTION, ALLOWABLES, NOMINATIONS, AND ESTIMATED CONSUMPTION IN TEXAS, 1961–62

(daily average in barrels)

Month and year	Allowables[a]	Actual production	Purchaser nominations	Bureau of Mines estimated consumption[b]
1961				
January	2,799,157	2,491,846	2,593,971	2,710,000
February	2,804,477	2,481,122	2,655,597	2,700,000
March	2,980,842	2,655,898	2,678,265	2,560,000
April	2,882,021	2,556,530	2,567,594	2,490,000
May	2,636,562	2,365,953	2,462,998	2,525,000
June	2,702,699	2,394,816	2,479,545	2,540,000
July	2,629,959	2,352,563	2,417,424	2,555,000
August	2,669,128	2,370,840	2,435,671	2,575,000
September	2,699,540	2,390,355	2,484,313	2,540,000
October	2,683,214	2,386,049	2,457,565	2,460,000
November	2,706,095	2,431,905	2,511,131	2,495,000
December	2,772,304	2,540,899	2,529,017	2,545,000
1962				
January	2,854,898	2,524,781	2,606,470	2,665,000
February	2,867,437	2,540,457	2,623,451	2,605,000
March	2,681,003	2,429,553	2,548,221	2,605,000
April	2,769,913	2,472,562	2,494,184	2,500,000
May	2,710,025	2,429,968	2,455,787	2,515,000
June	2,785,724	2,462,882	2,507,201	2,540,000
July	2,703,002	2,410,123	2,466,016	2,560,000
August	2,722,035	2,401,285	2,497,637	2,575,000
September	2,738,004	2,444,582	2,537,720	2,505,000
October	2,717,771	2,415,851	2,490,333	2,415,000
November	2,743,067	2,460,786	2,524,186	2,500,000
December	2,717,834	2,409,903	2,473,440	2,545,000

[a] The allowable figure for a month is the amount determined at the market-demand hearing the preceding month.

[b] U.S. Bureau of Mines estimates of consumption include some lease condensate while the other columns reported here do not.

SOURCE: Texas Railroad Commission, *Annual Reports of the Oil and Gas Division*, 1961 and 1962, Tables 5 and 14, and monthly reports on market-demand hearings.

This is done because experience has shown that production inter-ruptions and the inability of some wells to make their allowables cause actual production to fall below scheduled production.[17] Pur-chaser nominations may vary from actual production for several reasons. A nomination is calculated by a company in part by assum-ing a market-demand restriction factor that is applied to the fields in which it produces or purchases crude or receives crude through an exchange agreement. Normally a purchaser's nominations are made on the basis of crude that will accrue to his connections at a given percentage market-demand factor, and most purchasers when stating a percentage are not recommending a market-demand factor for the state as a whole. What is actually purchased may, therefore, vary from what is recommended. Among the thirteen major pur-chasers recommending a market-demand factor for November, 1963, one proposed 25 per cent, three proposed 26 per cent, five proposed 27 per cent, two proposed 28 per cent, and two proposed 29 per cent. The commission set a rate of 27.5 per cent.

A commission, after evaluating the information presented in the hearing, sets the state allowable in a general market-demand order. Of major importance in this determination is the commission's past experience and its "feel" for the situation. In some states the com-mission asks for the opinions of major purchasers to determine what each thinks the level of allowables should be to meet his par-ticular needs. This is an aid to getting a feel for the market. Rarely are all purchasers in exact agreement, and quite often there is sub-stantial disagreement. This is what would be expected, since each views the picture from his own particular production, stock, refin-ing, and market situation.

It is important to the commission to keep the allocated produc-tion closely in balance with forthcoming demand. Otherwise, it loses control over its most important function—that of allocating production back to individual wells and pools. If it chronically authorized production in excess of the amount demanded, the purchasers could pick and choose among the available sources, thus defeating the allocation purposes of the commission. The

---

[17] The relationship of allowables to production varies from state to state. For example, in Oklahoma during 1965 and 1966, production exceeded allowables in most months. In Texas, New Mexico, Kansas, and Louisiana (except in months when underproduction caused by a hurricane in an earlier month was being made up), allowables exceeded production.

initiative in the proration system would pass from the commission to the purchasers, into what is known as "purchaser proration." Since one of the primary purposes of proration is precisely to prevent purchaser discrimination among sources of supply, any large and continuing deficiency of this sort would undermine the system.

In setting a state total to be allocated to wells, each commission is acting for its own state alone. It is true, however, that the sum of the actions of all commissions covers a large fraction of the total domestic production of oil, and it has at times been suggested by outside observers that the whole system is a smokescreen to cover a monopoly structure in which purchasers, producers, and commissions are willing collaborators. Though a primary purpose and result of the system is to stabilize the market, the "conspiracy" view does not describe the facts. State commissions do not act collusively to "rig the market." Actually, each one operates separately in a fish bowl. Each knows what the other does, and the actions of each are to a significant degree interdependent. The system of communications is aided by the existence of the procedure of nominations by purchasers, since purchasers often nominate in several states. Thus, there is a situation in which each commission makes its decisions based in part on what other states do and in part on what purchasers in its own state do. Since purchasers live in a competitive market for products and crude, they must base their nomination decisions in part on what the commissions do in the states in which they purchase and in part on what other purchasers do in the aggregate. The system does not depend upon collusion among commissions or among purchasers or between a commission and the relevant purchasers. There is, however, substantial interdependency and a degree of parallel behavior based upon generally known market facts and aimed at maintaining market stability.

At later points, and especially in connection with prices, we shall have more to say about the market structure for crude oil.[18] At the present point, the nature of the proration system can be most clearly seen if, abstracting from the broader market picture, we look at it as a device by which each state attempts to serve its own separate interest. A commission authorizes as much production as it has reason to believe purchasers will take under existing market conditions. Existing market conditions implicitly include stable

[18] See Chapter 8.

prices. A commission cannot force more on unwilling purchasers without endangering the price structure;[19] it has no reason to authorize less. Its principal task, therefore, is to allocate the anticipated amount of production among the various wells in the state, which are in the aggregate capable of producing much more than can be sold.

DETERMINATION OF POOL AND WELL ALLOWABLES

Once the statewide allowable has been fixed in a market-demand state, the task becomes one of dividing this total among pools and specifying the portion of the pool allowable to be produced by each well. The actual procedure is not one of determining in sequence, first, the pool allowables, and second, the well allowables. The usual (but not universal) practice is to apply a formula that governs the production of wells in pools and that simultaneously determines the total allowable production from the pool.[20] The formulas, once established, can be administered on a routine basis, but their terms reflect compromise solutions among the varied and contradictory purposes of conservation regulation. The various state statutes specifically say or have been interpreted by the courts to mean that in allocating production among wells within a pool a commission has an obligation to prevent unnecessary waste and also an obligation to protect the individual rights of the mineral ownership interests. A less well-established principle, but one that is also generally recognized, is that a commission must not "unreasonably" discriminate among pools or fields in dividing up the state allowable.[21] A final overriding principle that has, to our knowledge, no statutory authority is to provide an atmosphere in which a high degree of individual initiative and enterprise is possible. This last principle is deeply ingrained in the regulatory authorities themselves, as well as in the industry and the legislatures in the respective states.

---

[19] This was attempted in Oklahoma and was struck down by the State Supreme Court. *Gulf Oil Corporation v. State of Oklahoma, et al.,* 369 P. 2d 933 (1961).

[20] See IOCC Governors' Special Study Committee, *op. cit.,* pp. 63–77, for a detailed discussion of allocation systems of the various states.

[21] North Dakota and New Mexico treat widely separated producing areas differently. Also, Louisiana and Texas treat offshore production differently from onshore production.

The formulas for allocating a state quota among pools and wells are as diverse as the number of market-demand states. There are three elements of similarity between them: (1) there is some restriction below maximum efficient capacity of at least some wells and pools in each state; (2) the bench mark or starting point for figuring cutbacks is usually tied to the depth of the production; (3) some adjustment in allowable production is made for differences in well spacing or size of the proration unit. Beyond these general similarities, the proration formulas are very different. We shall use Texas, Oklahoma, and Louisiana to illustrate similarities and differences.

"Yardstick" or "Top" Allowables

The basic proration tool in all market-demand states is the depth-factor, or "yardstick," schedule, adjusted by an acreage factor.[22] In its essentials, such a schedule may be described as follows: Every pool is produced at some particular depth. According to its depth, each well in a pool is assigned a "top" or "schedule" allowable that represents a hypothetical 100 per cent rate of production if it were not restricted, the amount increasing with the depth of the pool. The actual, or permitted, rate of production can then be expressed as a percentage of the base figure. At any particular depth, the size of the base figure is adjusted to the acreage assigned to each well as a production or drilling unit.

The depth-factor schedules for Texas, Oklahoma, and Louisiana are shown in Table 8. Part 1 of the table shows the 1947 Yardstick and Part 2 the 1965 Yardstick for Texas. The 1947 Yardstick applies to fields discovered prior to January 1, 1965. The 1965 Yardstick applies to fields discovered after this date.[23] For Louisiana (Part 4 of Table 8) only the March, 1953, schedule is shown, which applies to fields discovered since that time. There is another schedule for fields discovered earlier. As we noted above, the depth-factor sched-

---

[22] Depth-factor schedules are given different labels in different states, but they all involve the same general principle. In Texas the depth-factor schedule is called the 1965 (or 1947) allowable yardstick schedule, in Oklahoma the top-per-well allowable schedule, and in Louisiana the depth-bracket allowable schedule. New Mexico and Kansas have similar systems with different nomenclature.

[23] Texas Railroad Commission, Oil and Gas Division, Order No. 20–54, 115, November 20, 1964.

ules for the three states indicate that allowables vary not only with depth but also with the size of the production unit. It is also apparent that allowables are not proportionate to acreage under these schedules. In Texas, for example, under the 1947 Yardstick the basic allowable for any depth bracket is on the 10-acre unit. For larger units one barrel of allowable is added for each acre over the basic ten. Thus a 7,900-foot well on 10 acres has a top allowable of 91 barrels, and, if it is on 80 acres, 161 barrels (91 barrels + [1 barrel × 70 acres]). Such a system puts considerable emphasis on the well factor or size of the unit. We shall return to this point shortly.

The adoption of the 1965 Texas Yardstick was a result of a realization on the part of the Texas commission and the industry that the number of producing wells in Texas was rising while the level of production and reserves was remaining relatively static. Commissioner William Murray in a speech in 1962 announced the commission's plans to revise the Yardstick and made it clear that the revision was aimed at reducing the number of development wells drilled in any field. He was reported as stating that "... we are doubling the reserves which an oil well will recover when we double spacing. But the barrel-an-acre provision of the present Yardstick adds only a small percent more allowable. And the economic incentive is for closer spacing." [24]

The commission appointed an industry-advisory committee to study the situation and come up with a revision. In its statement at a commission hearing the committee noted that, "A principal effort has been directed toward the relationship of the allowables assigned at various spacings to determine whether the acreage credit is adequate to prevent the drilling of unnecessary wells. ... Assumptions have been made based on industry experience of costs under varying factors of reserves, depth, producing capacities, market demand percentages, and income tax brackets to determine their effect in obtaining the objectives of revised allocation." [25]

The committee presentation clearly illustrates the inducements for overdrilling because of inadequate acreage credits. It also pointed out that, "Under inadequate allocation yardsticks, any improve-

---

24 *Oil and Gas Journal,* October 29, 1962, p. 119.

25 Prepared statement by the Texas Oil Advisory Committee before the Texas Railroad Commission, hearing date July 16, 1964, p. 1.

TABLE 8.   DEPTH-FACTOR SCHEDULES FOR
TEXAS, OKLAHOMA, AND LOUISIANA

Part 1: 1947 Texas Yardstick and Marginal-Well Allowables

(barrels daily per well)

Depth bracket (feet)	Yardstick			Marginal-well allowable
	10-acre	20-acre	40-acre	
0– 1,000	18	28	–	10
1,000– 1,500	27	37	57	10
1,500– 2,000	36	46	66	10
2,000– 3,000	45	55	75	20
3,000– 4,000	54	64	84	20
4,000– 5,000	63	73	93	25
5,000– 6,000	72	82	102	25
6,000– 7,000	81	91	111	30
7,000– 8,000	91	101	121	30
8,000– 8,500	103	113	133	35
8,500– 9,000	112	122	142	35
9,000– 9,500	127	137	157	35
9,500–10,000	152	162	182	35
10,000–10,500	190	210	230	35
10,500–11,000	–	225	245	35
11,000–11,500	–	255	275	35
11,500–12,000	–	290	310	35
12,000–12,500	–	330	350	35
12,500–13,000	–	375	395	35
13,000–13,500	–	425	445	35
13,500–14,000	–	480	500	35
14,000–14,500	–	540	560	35

SOURCE: Texas Railroad Commission, Oil and Gas Division.

ments in prices or market demand factors will increase close spacing incentives and stimulate the development of excess producing capacity." Also under such yardsticks ". . . lower operating costs . . . [are] apt to lead to close spacing."[26]

The revised Yardstick establishes the 40-acre unit as basic. For 10- and 20-acre spacing the allowables are on a 100 per cent acreage basis except for shallow wells, i.e., 20-acre spacing gets one-half the 40-acre allowable. For 80- and 160-acre spacing a variable acreage

[26] *Ibid.*, p. 6.

TABLE 8 (continued)

Part 2: 1965 Texas Yardstick and Discovery Allowables[a]

(barrels daily per well)

Depth bracket (feet)	Yardstick					Discovery allowable
	10-acre	20-acre	40-acre	80-acre	160-acre	
0– 2,000	21	39	74	129	238	20
2,000– 3,000	22	41	78	135	249	60
3,000– 4,000	23	44	84	144	265	80
4,000– 5,000	24	48	93	158	288	100
5,000– 6,000	26	52	102	171	310	120
6,000– 7,000	28	57	111	184	331	140
7,000– 8,000	31	62	121	198	353	160
8,000– 8,500	34	68	133	215	380	180
8,500– 9,000	36	74	142	229	402	180
9,000– 9,500	40	81	157	250	435	200
9,500–10,000	43	88	172	272	471	200
10,000–10,500	48	96	192	300	515	210
10,500–11,000	–	106	212	329	562	225
11,000–11,500	–	119	237	365	621	255
11,500–12,000	–	131	262	401	679	290
12,000–12,500	–	144	287	436	735	330
12,500–13,000	–	156	312	471	789	375
13,000–13,500	–	169	337	506	843	425
13,500–14,000	–	181	362	543	905	480
14,000–14,500	–	200	400	600	1,000	540

[a] Marginal allowables are the same as those shown in Part 1 of this table.
SOURCE: Texas Railroad Commission, Oil and Gas Division, Order No. 20–54, 115, November 20, 1964.

credit is used, starting at 75 per cent at the shallowest depths and gradually diminishing to 50 per cent. Conversations with industry people indicated that comparative rate-of-return studies were run to determine the attractiveness of producing under the revised yardstick at a 28 per cent market-demand factor as compared to producing under the "new-pool" depth schedule in Louisiana under a 33 per cent market-demand factor. There was an obvious concern with making the Texas system more "competitive," i.e., more attractive as an area for exploration and development.

In Oklahoma the same general patterns are found. (See Part 3 of Table 8.) Allowables increase with the spacing pattern for any given depth although the incentives for wider spacing are lacking.

TABLE 8 (continued)

Part 3: Oklahoma Top-per-Well and Discovery Allowables[a]

(barrels daily per well)

Depth bracket (feet)	Regular top-per-well allowable				Discovery allowable[b]	Days to produce after discovery
	10 acres or less	20 acres	40 acres	80 acres		
0– 2,000	25	42	53	–	30[c]	270
2,801– 3,000	30	45	57	–	36	340
3,801– 4,000	35	49	61	–	42	410
4,801– 5,000	40	53	65	79	48	480
5,801– 6,000	48	60	75	94	56	570
6,801– 7,000	56	70	88	110	68	660
7,801– 8,000	65	81	101	126	80	770
8,801– 9,000	76	95	119	149	92	880
9,801–10,000	90	113	141	176	108	1,000
10,801–11,000	115	144	180	225	138	1,030
11,801–12,000	153	191	239	299	184	1,050
12,801–13,000	203	254	318	398	244	1,050
13,801–14,000	253	316	395	494	304	1,050
14,801–15,000	303	379	474	593	364	1,050

[a] The Oklahoma schedules actually give figures for every 200 feet of depth. We have taken only 1,000-foot intervals and have shown the allowable for the 200-foot interval each 1,000 feet.

[b] Discovery allowables were raised from the regular 10-acre schedule to 120 per cent of the 10-acre schedule in June, 1966.

[c] There are different discovery allowables for each 200 feet between 1,000 and 2,000 feet.

SOURCE: Corporation Commission of Oklahoma, Conservation Division.

Here, the 10-acre schedule provides the basis for determining allowables on other spacing. With minor deviations, the 10-acre allowable is multiplied by 125 per cent to get the 20-acre allowable; the 20-acre allowable is multiplied by 125 per cent to get the 40-acre allowable; and the 40-acre allowable is multiplied by 125 per cent to get the 80-acre allowable. Thus, at 7,900 feet, the top allowable is 65 barrels per day on 10-acre spacing, 81 barrels on 20 acres, 101 barrels on 40 acres, and 126 barrels on 80 acres. The importance of well spacing in the fixing of allowables may be shown by a calculation from the preceding example. In the hypothetical field at 7,900 feet, if developed on 10-acre spacing, each 80 acres would have a top allowable of 520 barrels; on 20-acre spacing, 324 barrels;

TABLE 8   (continued)

Part 4: Louisiana Depth-Bracket Allowables, March, 1953

(barrels daily per well)

Depth bracket (feet)	40-acre spacing		80-acre spacing	
	Onshore	Offshore	Onshore	Offshore
0– 2,000	80	193	120	233
2,000– 3,000	95	214	143	262
3,000– 4,000	114	238	171	295
4,000– 5,000	134	265	201	332
5,000– 6,000	159	296	239	376
6,000– 7,000	186	331	279	424
7,000– 8,000	214	379	321	486
8,000– 9,000	239	416	359	536
9,000–10,000	274	463	411	600
10,000–11,000	310	512	465	667
11,000–12,000	347	559	521	733
12,000–13,000	383	605	575	797
13,000–14,000	431	668	647	884
14,000–15,000	483	734	725	976
15,000–16,000	557	830	836	1,109
16,000–17,000	645	942	968	1,265
17,000–18,000	726	1,053	1,089	1,416
18,000–19,000	816	1,167	1,224	1,575
19,000–20,000	927	1,307	1,391	1,771
20,000–21,000	1,057	1,469	1,586	1,998

SOURCE: Louisiana Conservation Department.

on 40-acre spacing, 202 barrels; on 80-acre spacing, 126 barrels. Thus, identical pools can have wholly different total allowables according to the spacing patterns. In the Oklahoma system, deeper production is generally given a larger increment of allowable for wider spacing than shallow production. The well factor still is important.

In Louisiana, the pattern is somewhat different. The depth-acreage schedule in Part 4 of Table 8 was adopted in March, 1953, and still serves as the basis for determining allowables in the state. It is, however, applied in two different ways. In 1960, after extensive hearings on well spacing and allowables, the Department of Conservation issued an order establishing procedures for "new" and "old" pools in the state.[27] The order was felt necessary in order, among other things, "to encourage wider spacing and uniform drilling patterns, to prevent the drilling of unnecessary wells, to effect a more orderly distribution of allowables, to economically and efficiently drain all pools. . . ." Old pools are those penetrated prior to May 24, 1960. In the case of old pools, the pool allowable is determined by multiplying the number of wells in the pool times the appropriate depth-bracket base allowable in the 40-acre column and adjusting for the market-demand factor. The pool allowable is then apportioned among wells in the pool according to productive surface acreage for each unit well. While the well-spacing regulation generally covering old pools provides distance requirements that establish roughly 20-acre minimum units,[28] in actual practice spacing in old pools was more often 40 acres.

The newer Order 29-H increased the size of the minimum-sized unit to 40 acres, which received the basic allowable according to depth, adjusted for the market-demand factor. On smaller than 40-acre spacing a unit received a fraction of the base allowable with the fraction determined by a denominator of 40 and a numerator equal to the actual acreage in the unit—for example, 30 acres. For wider than 40-acre spacing a unit receives an allowable that is equal to the 40-acre depth-bracket schedule plus one-half times the 40-acre schedule allowable times a fraction, the numerator of which is the productive acreage minus 40, and the denominator of which is 40. Strangely enough, Louisiana places a maximum of 90 productive acres that can be used in the numerator of the fraction, so that even though the drilling pattern may be on 160 acres, for allowable calculations a well can receive a maximum of $5/4$ths $\left(\dfrac{90-40}{40}\right)$ of $1/2$ of the 40-acre allowable. Thus the allowable for 80-acre spacing under the "new pool" regulations is 150 per cent of the 40-acre one. The pool allowable, thus determined by summing the well allowables, is then apportioned to wells according to their productive acreage. Under Order 29-H the most common size unit has been 80 acres. The new pool offshore allowable for greater than 40 acres is the *offshore* 40-acre allowable plus one-half times the 40-acre *onshore* allowable times the fraction as defined above. As is

---

[27] Louisiana Department of Conservation, Statewide Order No. 29–H, November 10, 1960.

[28] *Ibid.*, Statewide Order No. 29–E, July 15, 1957.

true for onshore wells, no more than 90 productive acres can be used for computing the numerator of the fraction.

The Louisiana system for new pools does not go as far as the 1965 Texas Yardstick in encouraging wider spacing. Conversations with industry representatives indicate that, for many of the better fields in Louisiana, the 50 per cent increment for drilling on 80 rather than 40 acres is insufficient incentive to induce 80-acre spacing, and development has been on 40 acres. There is no incentive to go beyond 90 acres, especially in light of Louisiana's policy of making virtually no exceptions to its general rules.

As far as we know, the only state that sets allowables strictly on an acreage basis within each depth bracket is North Dakota.

The "top" or "yardstick" allowables that we have been describing are, it must be repeated, simply hypothetical 100 per cent base figures from which actual current allowables are calculated by applying a percentage "market-demand" factor. They do not in any sense represent the efficient productive capacity of the pools and wells to which they are applied. There is, indeed, a certain mystery about the rationale and construction of these diverse depth-factor schedules which we shall want to examine a little later.

A depth-factor schedule creates a series of proration formulas that vary with depth and well spacing. This can be illustrated using the 1947 Texas Yardstick schedule. If we assume that the 10-acre schedule gives a 100 per cent well-factor proration formula for any depth, it is then possible to compute the relative weights of the well factor and the acreage factor for other than 10-acre patterns at any depth. Table 9 illustrates the change in relative weights for selected depths using 40- and 80-acre spacing.[29]

Thus the 1947 Texas Yardstick gives a 47 per cent well—53 per cent acreage formula for 1,000- to 1,500-foot wells on 40-acre spacing, but gives an 84 per cent well—16 per cent acreage formula for 9,500- to 10,000-foot wells on the same spacing. The same sort of example could be worked out for Oklahoma, and the same general patterns would appear. If 10-acre spacing allowables are computed for Louisiana and these taken as 100 per cent well situations, the well factor

---

[29] For an interesting discussion of this and other aspects of Texas regulation, see Granville Dutton, "Proration in Texas—Where Is It Heading?" a paper presented to the Southwestern District, Division of Production, American Petroleum Institute, Midland, Texas, March 19, 1964.

TABLE 9. RELATIVE IMPORTANCE OF WELL AND
ACREAGE FACTORS AT VARIOUS DEPTHS IN THE
1947 TEXAS YARDSTICK

Depth bracket (feet)	10-acre allowable (barrels)	Relative importance of well and acreage factors			
		40-acre spacing		80-acre spacing	
		% well	% acreage	% well	% acreage
1,000– 1,500	27	47	53	28	72
3,000– 4,000	54	64	36	44	56
7,000– 8,000	91	75	25	57	43
9,500–10,000	152	84	16	68	32

for new pools on 80-acre spacing is only 17 per cent at 2,000 to
3,000 feet and 14 per cent at 7,000 to 8,000 feet. It is evident that
the 1965 Texas Yardstick gives much greater weight to acreage
throughout all depth ranges and spacing patterns than do the systems in Oklahoma and under the 1947 Texas Yardstick. The Louisiana system for new pools gives greater weight to acreage up to 90
acres, but the acreage factor rapidly gets smaller on wider than
90-acre spacing.

Taking fifteen hypothetical pools of 800 productive acres each,
let us make the following assumptions:

(1) Four fields in Oklahoma, one each on 20-acre, 40-acre, 80-acre,
and 160-acre spacing.

(2) Eight fields in Texas, four each with the same spacing as in
Oklahoma for the 1947 Yardstick and four for the 1965 Yardstick.

(3) Eight fields in onshore Louisiana, four each with the same
spacing for "old" pools and four for "new" pools.

(4) Wells in each case are drilled on equal spacing.

(5) Well depth in each case is 7,900 feet.

(6) Assume 100 per cent market-demand factor in each state.

(7) Wells in each case can produce their top allowables.

(8) Cost of all wells is roughly the same.

Table 10 indicates the variations in allowables among the different states and depth-acreage schedules. The results are rather

TABLE 10. COMPARISON OF DEPTH-ACREAGE SCHEDULES
AMONG STATES

Item	Oklahoma	1947 Texas	1965 Texas	Old Louisiana	New Louisiana
Acres	800	800	800	800	800
Depth-feet	7,900	7,900	7,900	7,900	7,900
Number of wells					
20-acre	40	40	40	40	40
40-acre	20	20	20	20	20
80-acre	10	10	10	10	10
160-acre	5	5	5	5	5
Well allowable					
(barrels per day):					
20-acre	81	101	62	214	107
40-acre	101	121	121	214	214
80-acre	126	161	198	214	321
160-acre	157	241	353	214	348
Pool allowable					
(barrels per day):					
20-acre	3,240	4,040	2,480	8,560	4,280
40-acre	2,020	2,420	2,420	4,280	4,280
80-acre	1,260	1,610	1,980	2,140	3,210
160-acre	785	1,205	1,765	1,070	1,740

startling. For any given spacing, the allowables in Oklahoma are always the lowest. In terms of encouraging extremely dense drilling, the systems for old pools in Louisiana, the 1947 Texas Yardstick, and Oklahoma are the worst offenders. The drastic change in Louisiana in 1960 created great disincentives for close spacing relative to what existed there in the past, although there are still disincentives to go above 90-acre spacing under the "new pool" rules in this state. The change in Texas from the 1947 to the 1965 Yardstick was appropriate in that almost no incentives were given for close spacing, which was quite the contrary under the 1947 version.

In order to make specific judgments about incentives, one must not only know allowables and the price per barrel produced, but also the cost of drilling, completing, and producing the wells. If we assume that all these wells have the same costs, it is obvious that in Texas, for example, the shift from the 1947 to the 1965 Yardstick creates a situation in which there are only 60 more barrels of allowable on 20-acre spacing than on 40-acre spacing, but the number of wells and hence costs have doubled. Whether it pays to go to 80-acre or 160-acre spacing under the 1965 Yardstick depends on well costs. With the halving of costs on 80-acre spacing, it may pay to go this route, since revenues have dropped only about 18 per cent (disregarding the interest cost of delayed payout) as compared with revenue from 40-acre spacing. We should emphasize strongly that we are ignoring in the discussion the market-demand factor which also varies among these three states and thus also influences production and incentives. For the past several years, the market-demand restriction has been somewhat less in Louisiana than in Oklahoma or Texas, thus increasing incentives in Louisiana relative to the other two states.

It is perhaps safe to say that under the Louisiana system for new wells and the 1965 Texas Yardstick there will be wider spacing, at least up to 80 acres, than was found under the older systems in the two states.[30] Over all, however, general conclusions are hard to establish. The differences in regulations among states have no rational basis for the simple reason that depth-factor schedules adjusted for spacing patterns also have no rational theoretical basis. All are arbitrary systems in the sense that they have no firm and continuing bases in fact.[31] It is possible to say something about the effects of one system as opposed to another under given conditions, but it is not possible to generalize and proclaim one or another as conforming more or less closely to a recognized norm.

This highly simplified discussion of the complex system of allowables and spacing is sufficient to point up how difficult it is to get any real picture of what domestic oil costs are, or what they would be if reservoirs were efficiently developed and produced. The cost per barrel of production obviously varies greatly among our twenty hypothetical fields. On what grounds are these differences justified? Those in government, industry, and elsewhere who are concerned with the costs of national security in the oil industry must, of necessity, be concerned with the production costs of domestic oil. Certainly the consuming public has an interest. Are these costs as low as possible? We shall return to this question at a later time.

---

[30] It should be remembered that spacing regulations and allowables for new wells in old pools are governed by the existing rules in that field. Thus, much development drilling may yet be done under the older systems.

[31] This statement should be qualified in that the 1965 Texas Yardstick was devised by estimating market-demand factors, costs, reserves, discount rates, etc. Whether it will be revised as these factors change remains to be seen.

We may now make one further point by relaxing the assumption that all the wells in our hypothetical field are on units of equal size. The pool allowable under this assumption, it will be remembered, could be computed by merely multiplying the schedule allowable times the number of wells or production units. In our twenty-well (40-acre spacing) field, the pool allowable in Texas was, under both yardsticks, 2,420 barrels, or 121 barrels per well. Let us assume that the commission established a proration formula with a 25 per cent well factor and a 75 per cent acreage factor. This means that 30.25 barrels are attributable to a well and 90.75 barrels are attributable to the 40 acres per production unit, giving the total 121-barrel allowable. As long as wells are equally spaced, the well factor in the formula causes no problem. Let us now, however, assume that there are nineteen wells on 40-acre spacing and four wells on only 10 acres each. Each of the 10-acre wells receives an allowable of 30.25 barrels for the well factor plus 10/40 of 90.75 barrels, or 22.69 barrels, for the acreage factor, or a total of about 53 barrels. The pool allowable is the sum of the well allowables or 2,511 barrels (121 × 19 + 53 × 4). If the commission had established a formula based on 100 per cent acreage, the 10-acre wells would have received an allowable of one-fourth of 121 barrels or 30.25 barrels, rather than the 53 barrels under the 25–75 formula.

This discussion is included here to clarify the pool-allowable concept. The above example shows that pool allowables are not automatically determined by the depth-factor schedules, but rather are dependent upon these schedules plus the proration formulas used in pools in which irregular spacing occurs. When a 100 per cent acreage formula is used, the pool allowable can be computed from the depth-factor schedule. In all other cases it is also necessary to apply the special proration formula. This has more significance in intrapool allocations and various problems of "fair shares," drainage, and correlative rights than it has for the purely mathematical determination of allocation to pools as such.

"Nonallocated" or "Exempt" Production

As a general rule the pools that are regulated under the depth-factor schedules are subject to market-demand restrictions, although there are some exceptions to this. The matching up of total production with estimated demand cannot, however, be done simply by applying a percentage cutback to the base schedules. The reason is that various classes of wells and/or pools are exempt from market-demand restrictions. Our three states illustrate three different ways of handling the exempt versus nonexempt production problems. Texas and Louisiana portray extreme situations—the former having several types of exempt production and the latter having no specifically exempt production. Oklahoma is in a somewhat intermediate position.

In Texas there are several types of wells or pools that get partial or full exemption from market-demand restrictions. The major categories include the following:

1. *Discovery wells* will be discussed below. These wells are put on a special depth-factor schedule and are exempt from market-demand restrictions for a limited period of time.[32]

2. *Statutory marginal wells* are defined as marginal by statute and are entirely exempt from production restrictions. A well is marginal if it is on artificial lift and if it produces no more than the following amounts of oil daily from the following depth brackets:[33]

Depth (feet)	Daily maximum allowed to be defined as marginal (barrels)
0 – 2,000	10
2,000 – 4,000	20
4,000 – 6,000	25
6,000 – 8,000	30
Over 8,000	35

Entire fields may be marginal, or individual wells may be so defined in fields that are subject to restrictions.[34]

3. *County regular fields* are a special category of fields that are given exempt status. In 25 north central Texas counties[35] there are numerous small, shallow, and relatively old pools. New drilling is, however, still going on in some of these areas. For proration purposes certain designated pools in each county are lumped together

---

[32] See p. 159 for a discussion of discovery wells.

[33] 102 Revised Civil Statutes of Texas, Article 6049b.

[34] For an interesting discussion of the marginal-well statute in Texas, see R. E. Hardwicke and M. K. Woodward, "Fair Share and the Small Tract in Texas," *Texas Law Review*, Vol. 41 (November, 1962), pp. 89–92.

[35] A majority of the counties in Texas Railroad Commission Districts 7B and 9.

in a "county regular field" for that particular county. These pools have their own depth schedule for allowables. It is as follows:

Class	Depth (feet)	Allowable (barrels per day)
1	0 – 2,000	16
2	2,000 – 3,000	28
3	3,000 – 5,000	35
4	below 5,000	40

This allowable schedule became effective January 1, 1950. No market-demand restrictions are applied to these wells. The 16-barrel-a-day allowable compares with 10 barrels daily for marginal wells shallower than 2,000 feet. On either the 1947 or 1965 Yardstick, with 20-acre spacing and a market-demand factor of 30 per cent, the allowed production for wells in this depth range would be less than 16 barrels daily. Generally speaking, discovery allowables do not apply to county regular wells.

4. *Capacity water-flood fields* are fields that are in the secondary recovery stage and in which water flooding is taking place. Such fields, even though they may not qualify for statutory marginal well status on the basis of production are often considered to be stripper fields and are often exempt from restrictions. Such classification requires a commission hearing and approval.

5. *Piercement salt-dome fields* are fields usually located on the flanks of salt domes that have pierced upward through overlying formations. The oil is often found in small pockets along the flanks. Presumably the justification for exemption of these fields is their smallness in area extent. Dutton reports that, in 1963, production from piercement salt-dome fields in Texas was about 98,000 barrels daily.[36] There are no rules in Texas that cover salt-dome field allowables generally. Each field is a special situation and is given a hearing. It is reasonable, however, to view the relatively shallow piercement-type salt-dome fields as an exempt or partially exempt category. Deep-seated domes are not usually given any special treatment.

There is also a miscellaneous group that includes various sorts of special cases of no great significance. There is sometimes a bonus allowable given for gas or water injected for pressure-maintenance

[36] Dutton, *op. cit.*, p. 10.

purposes, or for shut-in wells in unitized projects. This, however, is done on a field-by-field basis with no general rule applicable. Bonuses may be given for salt-water disposal also.

Exempt production in Texas is substantial and is the cause of some of the regulatory problems facing the state. In April, 1963, exempt production in Texas broke down in the following way:[37]

	(barrels daily)
Discovery wells	112,821
Statutory marginal wells	412,260
County regular, capacity water floods, others	688,772
Total exempt production	1,213,853

This exempt allowable of 1,213,853 barrels daily was out of a total state allowable of 2,818,812 barrels daily, or about 43 per cent. This has a great deal to do with the degree of market-demand restriction that is placed on nonexempt wells.

A fuller picture of exempt allowables appears in Table 11, which is taken from the Dutton article cited above. Between 1958 and 1963, exempt production rose fairly steadily, both in absolute terms and as a percentage of total Texas production. The growth of exempt production, coupled with a relatively stable demand (at going prices) for Texas oil, resulted in a decrease in the market-demand factor (an increase in the degree of restriction on prorated wells). It was Dutton's view that exempt production will level off during the 1964–68 period so that prorated production will increase to meet rising state demand.[38] However, even by 1968, he sees the market-demand factor as rising to no more than 33 per cent.[39] In May, 1966, the market-demand factor was 35 per cent, although at the time of writing it seems unlikely that this high a rate of production can be sustained.

Oklahoma classifies its pools, for proration purposes, into allocated, unallocated, pressure maintenance, secondary recovery, and special exempted pools. The allocated pools are those subject to market-demand restrictions. Our immediate concern is with the other categories.

---

[37] Records of the Texas Railroad Commission.

[38] It was estimated that exempt production in the spring of 1966 stood at 1.25 million barrels daily. *Oil and Gas Journal*, April 11, 1966, p. 43.

[39] Dutton, *op. cit.*, pp. 10–12.

TABLE 11. TEXAS ALLOWABLE DATA, YEARLY AVERAGES,
1958–63

(*thousands of barrels per day*)

Allowables

Year	Total	Exempt		Prorated		Schedule[a]	Market-demand factor
			(*per cent*)		(*per cent*)		(*per cent*)
1958	2,820	1,102	39.1	1,718	60.9	5,134	33.5
1959	2,932	1,144	39.0	1,788	61.0	5,315	33.6
1960	2,752	1,187	43.1	1,565	56.9	5,495	28.5
1961	2,766	1,211	43.8	1,555	56.2	5,612	27.7
1962	2,758	1,239	44.9	1,519	55.1	5,722	26.5
1963	2,843	1,237	43.5	1,606	56.5	5,742	28.0

[a] Scheduled allowables subject to market-demand factor.

SOURCE: Granville Dutton, "Proration in Texas—Where Is It Heading?" a paper presented to the Southwestern District, Division of Production, American Petroleum Institute, Midland, Texas, March 19, 1964. Data taken from Texas Railroad Commission records.

1. *Discovery wells,* while not included in the above grouping, constitute a type of exempt well. The general rule provides for a new field to be produced at 120 per cent of the regular 10-acre schedule for a specified period of time without market-demand restriction. This schedule is included in Table 8, Part 3.

2. *Unallocated pools* are a large group of pools that the Oklahoma Commission has exempted from restrictions. The rule governing whether a pool is put in this category is quite general and reads, in part, as follows:

All wells, common sources of supply, and areas not shown to require specific regulation ... for the prevention of waste, restriction of production to market demand, or from which the ultimate recovery from the reservoir will not be materially affected, and will not adversely affect the correlative rights, shall be classified as unallocated wells, common sources of supply, or areas.[40]

This leaves it to the discretion of the commission to exempt a pool or not. There is one restriction on unallocated pool allowables and that is that the maximum per-well-allowable shall not be greater than the basic minimum allowable for allocated pools as

[40] Oklahoma Corporation Commission, Conservation Division, Rule 303–2–A.

restricted by market demand. This means that if the smallest allowable for allocated wells is 8 barrels daily per well, then unallocated
wells can produce no more than 8 barrels daily. If the market-
demand factor reduces the lowest allocated well allowable to 7 barrels, this then becomes the maximum for unallocated wells.

3. *Water-flood projects* for secondary-recovery purposes are allowed to produce 20 barrels daily per well. This is varied by the
commission depending on specific circumstances. There are a few
other miscellaneous special exemptions provided, but none of great
significance.

In Oklahoma about 55 to 60 per cent of 1965 production fell
in the unallocated category. As in the case of Texas, the amount
of exempt production has a substantial impact on the degree of
restriction placed on nonexempt wells.

Louisiana has chosen to handle its "marginal" or less-than-flush
production in a different fashion from Texas or Oklahoma. There
the system designates no special categories of wells, but rather applies the depth-factor schedule with modifications. The general
rule is that a well can produce to capacity so long as capacity is
less than the current market-demand restricted allowable in its
depth-acreage bracket. The following example illustrates the principle. Let us assume we have a 9,500-foot pool with 40-acre production units, and assume also that the market-demand factor is 33 per
cent. Under these conditions a well could produce 90 barrels daily.
If, because of natural decline, a well fell to 85 barrels daily in a
given month, it would be allowed to produce to capacity. The following month this well would be included as having a daily allowable of 85 barrels. If the market-demand factor fell to 30 per cent
and the allowable were cut from 90 to 82 barrels, this well could
produce no more than the 82-barrel allowable. However, it can
produce to capacity so long as it stays within the restricted allowable of that well's particular depth bracket.

Thus, while Louisiana does not define or specifically regulate
marginal wells, its depth-factor schedule has a built-in device that
discriminates among wells by depth. Our 85-barrel well can produce to capacity under a 33 per cent market-demand factor because
it is 9,500 feet deep. If it were 8,500 feet deep, it would have to
curtail production to 79 barrels daily. Obviously there are many
wells in Louisiana that make only a few barrels daily. They are, in
effect, exempt from regulation and their allowables calculated from

their capacity production. Market-demand restrictions do not affect their production until their depth-factor allowable is cut below their capacity. Louisiana regulations specifically exempt "stripper fields" and pools less than 3,000 feet deep from any restrictions created by the acreage factor in the depth-factor schedule.

Discovery Allowables

Different states follow different practices with respect to allowables for wells in newly discovered pools—some giving them preferred treatment, others holding them to the ordinary market-demand factor in their respective depth-acreage brackets.

In Texas, a discovery well is put on a special allowable schedule set forth in Table 8, Part 1. Until the spring of 1966 discovery allowables were granted to the first five wells completed in the field or for a period of 18 months from the completion of the initial well. In April, 1966, the time limit for discovery-allowable production was increased from 18 to 24 months, and in May, 1966, the number of wells was raised from five to ten. The statewide spacing rule of 40 acres applies in this early development period of the field. With the completion of an eleventh well or twenty-four months after the initial well is completed, the entire pool is usually put on the regular yardstick or depth-factor schedule. When a pool goes off its discovery allowable in Texas, the commission holds a hearing to establish special field rules. It is at this hearing that the spacing and allocation formulas are established, which may be more or less than 40 acres, as the facts justify. For some fields, something other than the depth-factor allowable is set. These, however, are exceptions to the rule and are not numerous.

During the period when the discovery allowable is in effect, these wells are not subject to any market-demand restriction. Thus, while the discovery allowable for a 7,900-foot well on 40 acres is 160 barrels daily compared with 121 barrels daily for the regular Yardstick, the 121 barrels are subject to market-demand restriction, currently about 28 per cent or about 34 barrels, while the 160 barrels are not restricted. It should be noted, however, that discovery allowables have no necessary relationship to efficient producing rates. A well may be capable of producing several times its discovery allowable. In this sense, therefore, some wells getting discovery allowables are only partially exempt from restrictions. On the other hand, some discovery wells may not be able to produce the full discovery allowable. In such cases they are allowed to produce to capacity until they go off the discovery allowable. One authority in the industry estimated that discovery fields usually produce about 70 per cent of their discovery allowable.

Oklahoma also makes provision for discovery allowables, but in a somewhat different manner. Oklahoma uses its regular schedule for 10-acre spacing as the discovery-allowable system but adds to this a schedule showing the number of days that discovery wells in each depth category are allowed to produce without market-demand restriction. In June, 1966, the discovery allowable was raised from 100 to 120 per cent of the 10-acre schedule. Any well completed within a new field during the prescribed exemption period can produce on the discovery-allowable schedule without restriction until the period on the initial well expires. Thus a well producing from 7,900 feet would have an allowable of 80 barrels per day for 770 days after the completion of the discovery well. A second well in the field, completed sixty days after the first well, would have an allowable of 80 barrels per day for 710 days, and so on. When the discovery exemption period has elapsed, the field goes on the regular depth-factor allowable schedule and is subject to market-demand restriction. The actual spacing in the field then governs the specific schedule to be followed.

Louisiana, our third example, makes no special provision for discovery allowables. It has only its regular depth-bracket allowable schedules, which are different for onshore and offshore wells. A field is put on one of the schedules as soon as it is discovered and is subjected to the same market-demand restriction as all other fields in the state. New Mexico also has no special discovery allowables. Kansas has a system similar in principle to that of Texas.

The rationale for creating a bonus allowable for discovery wells seems fairly clear. It is meant as an inducement to explore, and provides a way for the operator to recover fairly rapidly a part of his investment. If the hypothetical Oklahoma well we cited above had produced under the regular schedule, on 40-acre spacing with a 30 per cent market-demand factor, it would have produced 23,870 barrels of oil in 770 days (101 barrels × 30 per cent market-demand factor × 770 days). Under the discovery allowable of 80 barrels daily for 770 days it would have produced 61,600 barrels, if it were capable of doing so. The operator, of course, would be delighted to find the cash throw-off more than two and one-half times larger during the first twenty-six months of the well's life. Such discovery allowables are bankable also, and are attractive to financial institutions because of the rapid payout. Undoubtedly, many bank loans are made on discovery allowables that would not have been made under regular allowables. The differences in the Texas and Oklahoma systems for discovery allowables do not seem significant for our purposes. Both provide incentives. Whether one or the other provides greater incentive probably depends on the specific situation, such as depth, cost, spacing, and speed of drilling. It seems difficult to justify either system under a philosophy of waste prevention.

We can foresee one potential problem with the new Texas system of granting discovery allowables until 10 wells are drilled or 24 months have elapsed. If spacing rules are not adopted until the field goes off the discovery allowable, the spacing pattern has pretty well been determined by the drilling of the first 10 wells. There are clear incentives to space as closely as allowed (40 acres under the statewide rule), since each well gets a specific allowable regardless of spacing. Hopefully, the commission will anticipate this problem and impose wide spacing before 10 wells are completed in those situations which warrant wide spacing.

Louisiana has not felt the need to provide added incentives for exploration. Part of this is explained by the higher depth-bracket schedule that makes regular per well allowables fairly high relative to Texas and Oklahoma. For example, our 7,900-foot well, if it were in Louisiana onshore, would produce 71 barrels daily on 40-acre spacing at a 33 per cent market-demand factor, and would produce 126 barrels if offshore.

The procedure by which a pool is placed on a depth-factor schedule is about the same in all states. The regulatory authority supervises the testing of wells in a field, certifies the producing depth, gas–oil and/or water–oil ratios, and other indicators of reservoir performance. A hearing is usually held early in the development of the pool to establish field rules governing well spacing, the proration formula, casing programs, water pollution, and other relevant aspects of field operation. At this point in the history of the field several critical decisions are made. The size of proration units is of major importance since allowables are assigned according to production units. The size of the proration unit is usually the same as the drilling unit for the field. Thus we often find 10-,

20-, 40-, 80-, and 160-acre drilling and proration units, with a recent trend toward the larger units.[41]

Uniformity of size within a pool of both drilling and proration units is usually aimed at, though it is much more severely enforced in some states than others—Texas being well known in the past for its numerous exceptions in favor of small-tract drilling and its use of well and acreage factors in establishing allowables. Where there is uniformity in the size of drilling units within a field, and where drilling and proration units coincide, the Yardstick usually determines the allowable for the field. In fields that do not have a uniform drilling pattern, the Texas commission has in the past quite frequently used special proration formulas to establish the allowables for proration units, such as 50 per cent well and 50 per cent productive surface acreage, or 25 per cent well and 75 per cent productive surface acreage, or 100 per cent productive surface acreage. In Texas, the 50–50 and 25–75 formulas have been most common in the past, but recent court decisions may require them to be discontinued in favor of 100 per cent acreage or other formulas for protecting correlative rights based on net acre-feet, net recoverable oil, or other criteria. It is significant that, in 1962, about 69 per cent of the allocation formulas for new fields in Texas was based on 100 per cent acreage or acre-feet and only about 8.5 per cent on 50 per cent well and 50 per cent acreage. By 1966, the 50–50 formula had virtually disappeared. As recently as 1960, only 2 per cent of the new field allocation formulas called for 100 per cent acreage or acre-feet. The remainder included some recognition of the well factor.

Many fields in the past were drilled on an irregular pattern, and well factors were applied in establishing special proration formulas. These two aspects of Texas regulation have, perhaps more than anything else, been the subject of the greatest criticism both within and outside of the industry. Since 1961 the use of special proration formulas has dwindled, as we have shown above. Also, uniform spacing has become general in virtually all new fields. In other words, for recent and new fields, the Yardstick and uniform spacing

---

[41] A proration unit must be made up of acreage that can reasonably be expected to be productive. Obviously, no such requirement can be imposed on a drilling unit. Where a drilling unit contains unproductive acreage, the proration unit will usually be smaller and have an allowable corresponding to the number of productive acres.

have determined pool allowables. There are a great many fields that were developed years ago, and that operate under old rules.

Allocation of the State Allowable among Pools

With the amount of detail about proration systems that we have accumulated to this point, we can now turn back to the question of dividing a state allowable among wells and pools in the state.

In Texas, the various categories of exempt production are subtracted from the total state allowable. This leaves the amount of production to come from prorated wells. The commission knows, from its depth-factor schedule, proration formulas, and spacing patterns in prorated fields, the amount of oil theoretically corresponding to the Yardstick allowable for each field. It then is possible to apply a market-demand percentage figure to the theoretical 100 per cent depth-schedule production to arrive at a figure equal to the desired prorated production level. The appropriate percentage is communicated to all producers who know what their own base schedules are and can therefore apply the percentage to their own actual daily and monthly production. From there on, the problem is simply one of policing and making adjustments for overproduction and underproduction.

It should also be pointed out that the Texas law requires the commission to allocate the state allowable on a "fair and reasonable basis." The statute reads, in part, as follows:

In order to prevent unreasonable discrimination in favor of one pool as against another, and upon written complaint and proof of such discrimination, the Commission is authorized to allocate or apportion the allowable production of crude petroleum oil upon a fair and reasonable basis among the various pools in the State; provided, however, that in allocating or ascertaining the reasonable market demand for the whole State the reasonable market demand of one pool shall not be discriminated against in favor of any other pool; and provided further, that the Commission shall ascertain the reasonable market demand of each such respective pool as the basis for determining the allotments to be assigned each such respective pool, to the end that such discrimination may be prevented.[42]

There is an apparent conflict in the wording of the statute. The first part suggests that each pool should receive a fair portion of the total state allowable. The second part suggests that the demand

---

[42] 102 Revised Civil Statutes of Texas, Article 6049d, Sec. 6.

for each pool should be determined and allocation of the state total made on this basis. The courts in Texas have generally adopted the first approach, namely, that each pool should share in the total state demand. This requirement can and does create problems in interpool allocation. For example, during the Suez crisis in 1956–57, purchasers indicated a willingness to take more oil from Gulf Coast fields in Texas. However, the commission felt unable to raise allowables substantially in these fields because it could not, at the same time, raise allowables in other Texas fields, and particularly not in West Texas. It would have been impossible to move the additional oil out of West Texas because pipelines were full. Commissioner Thompson described this situation rather succinctly:

We had purchasers that wanted more crude, but they wanted it around the coast in selected fields.

And our rule is that it [allocation] must be straight across the board, every field must share ratably in the production.[43]

The Suez crisis merely made more acute a problem that is ever present in all market demand states. Competitive pressures force buyers to seek out the lowest cost sources of supply. Under a system in which all fields share in the state allowable, it is quite clear that these competitive forces are substantially blunted. We shall return to this point in a moment.

Roughly the same thing happens in Oklahoma. An estimate of the degree of restriction is necessary before the actual restrictions are set. This stems from the fact that unallocated pool allowables are affected somewhat by restrictions, although the total effect is not large. Given an estimate of unallocated production, this is subtracted from the total state allowable to determine how much prorated wells can produce. Oklahoma, as in the case of Texas, knows the theoretical amount these wells can produce under a 100 per cent depth-schedule allowable. A percentage is then applied to equate the two.

In Louisiana the commission, from its records, knows what wells produced at capacity under the previous month's restrictions, and how much this production amounted to. They can subtract this from the state allowable to get an amount for wells to be prorated. From

---

[43] Testimony of Ernest O. Thompson, member of the Texas Railroad Commission, at the hearings of the Committee on Interstate and Foreign Commerce, House of Representatives, 85th Cong., 1st Sess. (1957), *Petroleum Survey*, p. 209.

the depth schedule and spacing the theoretical 100 per cent production can be calculated and a percentage figure applied to this to arrive at allowable production from restricted wells.

It is now possible to step back and view the effects of the details described above. We find that prorated wells in the three states were producing in June, 1966, at the following rates:

	(per cent)
Texas	34.5
Oklahoma	38
Louisiana	36

The first thing to be said about these percentages is that they cannot be compared with one another as an indication of the relative degree of restriction in the various states. The reason is that the percentages apply to depth-bracket and acreage bases that are different for each state.

The second thing to be said is that the percentages do not express, even for single states, the degree of restriction as related to the efficient producing capacity, or MER, of prorated pools within the state. The reason for this is that the depth-bracket schedules are in a high degree arbitrary, and bear no relation to the efficient capacity of the pools and wells to which they are applied.

These two points are sufficiently explained by the account we have given of the depth-bracket basis of proration, but may be further illustrated from the case of Texas. As we noted earlier, in April, 1963, Texas had about 1,214,000 barrels per day of allowables from exempt sources, leaving 1,605,000 barrels from prorated sources. Since the latter was at a nominal rate of 28 per cent, this would give a nominal 100 per cent sum of yardstick allowables amounting to around 5,730,000 barrels per day. This is, however, a fictitious figure, since the Texas authorities estimate that in 1963 the wells in the prorated category had an efficient sustainable capacity of around 3,700,000 barrels per day. On this productive-capacity base, the current allowables on prorated wells would be about 43 percent rather than 28 per cent. We have no similar breakdown of figures for other states, but any similarity between top allowable schedules and MER ratings is bound to be purely coincidental.

The general proration procedures outlined above are subject to exceptions in all states. Special field rules may be set up. In Texas, for example, a few pools are prorated on what is called an MER basis, including the giant East Texas Field, which has special allowable rules. In some fields in which an efficient rate of sustainable production is shown to be greatly in excess of the 100 per cent yardstick allowables, the commission may use the MER rather than the depth schedule as the base upon which to apply the market-demand restriction. What criteria are used in making such exceptions and what technical methods are used for calculating MER's are points upon which we are not sufficiently informed to describe the procedures; the ad hoc use of the MER concept for proration purposes does not correspond at all closely to definitions of MER to be found in the scientific literature.

It is not difficult to see why in certain situations there should be great pressure for exceptions to the strict application of the arbitrary depth-bracket formulas. The wells in one pool at a certain depth may be incapable of producing their 100 per cent yardstick allowable; the wells in another pool in the same bracket may be capable of producing the yardstick many times over. Some recognition of these differences may be given by exceptions in favor of the more prolific pools.

Interpool allocation creates some of the greatest inefficiencies found under the current regulatory system. These problems have long been recognized by regulators and others, but proposals for what might be workable alternatives are few. One proposal made in 1964 contains several elements found in other alternative systems.

The statutory standard, "fair and reasonable," for distributing allowables among the various pools is not subject to technological definition. Although it seems fair to allocate the statewide demand among fields on the basis of the relative initial reserves in each pool, it is probable that such a method would result in some pool allowables being higher than the pool's maximum efficient rate (MER) and others being so low as to preclude continued economic production.

The obstacles could be overcome by assigning each pool an MER as the ceiling allowable and an economic allowable which would constitute a floor. Where distribution of the statewide market demand on a reserves basis exceeded either of these limits, the limit would be set as the allowable. The sum of the economic allowables and the allowables limited by MER ceilings would be subtracted from the total state market demand and the remainder distributed on a reserves basis among other fields within the state.[44]

---

[44] Dutton, *op. cit.*, p. 7.

There are obvious difficulties in such a proposal, since it requires the estimation of reserves, MER's, abandonment rates of production, an economic allowable, and the like. Also, one can argue whether reserves are a proper basis for sharing in the first place. It does have the merit of having *some* rationale for interpool allocation, which is more than can be said for the present systems.[45]

THE DEPTH BASIS FOR PRORATION FORMULAS

The preceding survey shows that, except in the case of exempted categories, allocation formulas in the market-demand states are all based upon depth factors, associated in greater or less degree with an acreage factor. This fact having been observed, several questions naturally arise: (1) Why should depth-factor schedules be the universal method? (2) How does a state establish a depth-factor schedule? (3) Why are states so divergent in their depth-factor schedules? (4) Why are the acreage-factor elements so divergent? These questions divide into the aspects of uniformity in principle and disparity in application. In approaching these questions, the negative observation may be made that no state bases its system of interpool allocation on the relative richness of the deposits with respect either to the amount of recoverable reserves in pools or the efficient rate of producibility.

There is a certain amount of mystery surrounding allowable schedules. There is no clearly defined geological answer to tell us that, in general, deeper pools are capable of producing efficiently at higher rates than shallower pools. It may be true that deeper formations are likely to have higher pressures and perhaps more gas, but such rules of thumb have so many exceptions to them that they are hardly suitable as the firm basis for establishing a depth-factor schedule. Nor is it generally held that deeper pools have larger accumulations of oil. There may be some slight correlation between depth and producibility but no one has put this forth seriously as a major factor in shaping allowables.

---

[45] The province of Alberta, Canada, in 1965 began assigning pool allowables on the basis of each pool's ultimate reserves less 50 per cent of its cumulative production. Certain minimum allowables were also established. This interesting plan is described in Alberta Oil and Gas Conservation Board, *Report and Decision on Review of Plan for Proration of Oil to Market Demand in Alberta* (Calgary, July, 1964).

The only rationale that we have heard for depth-bracket schedules is the economic one: deeper drilling and deeper production are more costly; greater incentives, in terms of higher allowables, are required to bring forth the exploration and development of deeper horizons. The following excerpts from the findings in a 1953 Kansas commission order point out some of this reasoning. The commission found:

That evidence was presented to the Commission concerning the advisability of amending Rule 82-2-109 ... to provide for increased allowables to deep wells in order to compensate to some extent for the additional cost of such wells and the risks incurred in exploring the deeper horizons; that evidence presented showed the comparative costs of drilling shallow and deep wells; that any well below 4500 feet is considered a deep well, and the deeper the well below that point, the greater is the cost per foot for the total depth; that the increased drilling cost is due, among other things to the following: (1) Larger size and gauge of casing which must be used; (2) Different class or grade of pipe which must be used to overcome the tensile forces in the string; and (3) Definite breaks in rates for drilling wells at depths of 4500 feet, 6500 feet and 9000 feet.

That according to a generally followed principle in the petroleum industry, the initial gross investment in an oil well (including only the out of pocket cost of drilling the well), should normally be returned in three years; that recovery of the initial investment on deep wells drilled in Meade and Seward Counties at the normal allowable would take five and one-third years.

That the Governor's Oil Advisory Committee, after extensive study of matters relating to deep producing wells, has recommended a percentage schedule allocating increased allowables to pools producing from deeper horizons proportionate to the increased comparative cost of drilling for the respective depths, as being most feasible based on cost of development, lifting cost and time element of the payout.

·   ·   ·   ·   ·   ·   ·   ·   ·   ·   ·   ·   ·   ·

That it appears adoption of the recommended schedule of allowables for deep wells would tend to materially encourage further exploration and greater development of the deeper horizons of Kansas, especially in the southwestern part of the state, thus aiding the conservation of oil resources of the state; and that it will not prejudice or react to the disadvantage of the shallow well producers or tend to discourage exploration and development of shallow horizons.[46]

Accepting for the moment the general rule of rising costs with greater depths, we find a proration system that has at least a lim-

---

[46] Corporation Commission of Kansas, Conservation Division, Docket No. 44, 344–C (C–3266), January 16, 1953.

ited rationale of allocating production on the basis of costs, that is, rewarding higher cost wells with higher allowables. Stripped to its essentials, this is what a depth-factor schedule provides in some rough fashion. The question that then arises is what incentives are required in the depth schedules to stimulate the deeper drilling. In a statement before the Railroad Commission, the Texas Oil Industry Advisory Committee advocated to the commission a revision of the 1947 Yardstick and pointed out that "... the extremely high allowables at the greater depths are more than compensatory for increases in drilling cost. The payout period under the 1947 Yardstick at 14,500 feet is estimated to be only about 60 per cent of that at 9,500 feet (2.8 vs. 4.9 years).... Allowables under the 1947 Yardstick are unnecessarily discriminatory in favor of deep drilling."[47]

In considering the effect of bonus payments for higher cost production at greater depths, it is necessary to point out that the depth factor is only one of the factors involved. The other is the acreage factor. Here, the authorization of more wells than is necessary for efficient drainage of the pool has the perverse effect of magnifying the costs and validating them by higher pool allowables.

Granting, however, that deeper pools should have higher allowables, the question then arises, how far can this logic be carried? The answer appears to be: not very far. It is generally true that in any given geographic area drilling costs per foot of hole increase with depth. It is not necessarily true that two 8,000-foot wells that are widely separated, say, one in West Texas and one on the Texas Gulf Coast, will cost approximately the same. To the extent, then, that cost factors enter into the depth schedules as incentive factors, it is on a very rough-and-ready basis of an averaging sort, possibly calculated with some care at the time the schedules were drawn, but necessarily diverging further and further from any factual basis as time goes on.

We have been unable to uncover any official statements in the three states under discussion explaining the origin or logic of existing depth-factor schedules, but conversations with industry people indicate that, at least in Texas, the 1947 Yardstick was designed to

---

take into account differences in drilling costs at different depths. As was noted in the quotation above, the 1965 Yardstick had elements of this also. Both yardsticks attempt to provide increases in allowables that were roughly proportionate to increases in costs at the time each was adopted.[48] It was not claimed that the absolute level of costs would be covered by allowables at any given depth. Our inquiries suggest that the same was true of the 1953 Louisiana schedule at the time it was adopted. Our informants tell us that representatives of the industry were called in for consultation and that the schedule represented a pooling of informed judgment and compromise of interests. As far as we know, no state has ever attempted to bring its schedules up to date in terms of changes in drilling costs in new and old environments. Oklahoma reviewed its schedule in 1961–62, but made no basic changes in the depth schedule. The changes in Texas in 1965 were more substantial. It is easily seen why, once a schedule is in operation, changes would meet severe resistance, since operations under the schedule create a network of vested interests.

It should also be noted that proportionate cost changes over time at all depths create no particular problem in terms of treatment in the allowable system. Problems are apt to arise if costs of drilling to one depth change relatively more than those for drilling to a shallower or deeper depth. It may be, for example, that the 1947 Texas Yardstick accurately reflected *relative* costs at different depths. If technological improvements were successful in reducing real costs of deeper wells, then the reduced allowables for deeper production in the 1965 Yardstick may be justified. They may merely reflect changes in relative costs. We do not know whether this is the case or not.

The rough-and-ready character of the cost basis is further suggested by the comparison of different states. A cursory check of the relative growth rates of the schedules in different states reveals wide differences. If we take at random the increase in allowables between 2,000 feet and 9,501 feet on 40-acre spacing, we find the relationships shown below.

If schedules were based on carefully computed cost averages, one

---

[48] The rate of increase with depth of the 1947 Yardstick takes a peculiar dip at 10,000 feet. This was the result of an amendment to the 1947 Yardstick to assign allowables below 9,500 feet.

State	Barrels of daily schedule allowable		Increase in allowable	
	(At 2,000 ft.)	(At 9,501 ft.)	(Barrels)	(Percentage)
Texas: 1947..........	75	182	107	143
Texas: 1965..........	78	172	94	121
Louisiana...........	95	274	179	188
Oklahoma..........	53	131	78	147

might expect to find, if not roughly equivalent allowables at the same depth, at least schedules showing similar proportionate increase in allowables for given increases in depth. Nothing of this sort shows up. It is true, of course, that the schedules were drawn up at different dates, but the degree of inflation in drilling costs between 1947–48 (Texas) and 1953 (Louisiana) cannot account for the difference in the allowables of the two states. All in all, one is hard pressed to come up with reasons for the patterns of depth-factor schedules of any one state or for the difference among states. We have heard such things mentioned as competition among states for new development, the need to treat existing production fairly, and so forth. Such reasons are difficult to pin down.

As measuring an incentive structure, the top-allowable schedules are strictly fictitious, since no one is permitted to produce at the nominal rates, but only at some fraction of them. For example, Louisiana allowables in 1965 ran at about 34 per cent of the schedule rate. All the current incentives for new drilling are expressed in the 34 per cent column. It would make no difference if the 100 per cent schedule of 1953 were abandoned, and all future calculations based on the 34 per cent column, revalued to 100 per cent. The same is true for Texas, which in 1965 ran at a rate of about 29 per cent.

The importance of this aspect of the schedules was emphasized in the industry presentation on the 1965 Texas Yardstick before the commission. In its calculations of costs, revenues, and profitability of various hypothetical fields with assumed reserves and well spacing, the industry used a 28 per cent market-demand factor primarily, although some comparisons were made with a 40 per cent market-demand factor. The incentives for closer spacing under both the 1947 and 1965 Yardsticks increase as the market-demand factor rises. The 1965 Yardstick was based primarily on 1963–64 conditions, including the degree of market-demand restriction. Thus, when and if the market-demand factor changes, even assuming such other things as costs, prices, and discount rates do not change, the spacing and depth incentive patterns will change also. Since usually the market-demand factors in all states change in the same direction at the same time, movements in national demand also change the relationships of incentives among the relevant states.

In suggesting a "mystery" connected with the structure of depth schedules, we do not wish to suggest any mystery about why something of the sort almost necessarily exists. A depth-factor schedule for allowables serves primarily as an administrative convenience: a point from which to cut back production to market demand; a system by which everyone knows *where* he stands but not *why* he stands where he does relative to others. Without some such formula, commissions would have to take every reservoir as a special problem and, from cost data and other relevant factors, determine the terms of its participation—a forbidding administrative assignment. With respect to the whole subject, our own curiosity was aroused by the complete absence of any discussion of the arbitrary formulas that determine the position of every producer in the allowables hierarchy. Perhaps the principle at work is that of "sleeping dogs." If you have a system you can live with, why stir it up when you don't know what the alternatives would be? We feel that the entire system of depth-acreage schedules should be examined on its principles.

We can think of only two alternatives for which a clear rationale could be stated: (1) a modification of the present formulas with much more careful attention to the cost and other factors that enter into the formula; or (2) an entirely different system based upon MER ratings regardless of depth; or possibly (3) a combination of (1) and (2). To operate a market-demand proration system, it is necessary to have *some* base from which to calculate production allowables. What seems odd about the present system is the highly arbitrary character of the base.

In this discussion we have largely abstracted from the acreage factor that is associated with the depth factor in actual formulas. This introduces a different set of considerations, which are discussed elsewhere. There is, however, one important aspect of depth formulas that cannot be discussed without introducing the acreage factor. If depth formulas are designed to recognize the greater costs of deeper drilling, then their general rationale is that of providing incentives to such drilling. An interesting question then arises. Take a state like Texas where the formula favors the well factor as against the acreage factor, and increasingly so as depth increases. As we noted earlier, although the 1965 Yardstick is an improvement over the 1947 version, it could be argued that the effect of the formula is to encourage the drilling of *too many* wells at all depths, especially at greater depths, and especially as the market-demand factor rises. This question has to be associated with special incentive schedules for discovery wells. If the incentives for exploratory drilling at deep levels can be provided by special allowables, why is it necessary to provide incentives in the regular formula for drilling more development wells than are necessary for efficient drainage of a pool?

We shall leave these questions hanging at this point. Obviously, they reach out to a much wider range than the limited subject of depth schedules. They suggest, however, that the cost rationale for depth schedules cannot be considered by itself, in isolation from other considerations related to the subject of unnecessary development drilling—a subject dealt with at other points in this study.

ESTABLISHMENT OF PRORATION AND DRILLING UNITS

The preceding discussion has dealt primarily with the mechanics of establishing total state allowables and allocating state allowables among wells and pools. At several places we noted that well spacing or, more precisely, the size of the production or proration unit influences well allowables and thus pool allowables. This arises from the fact that all states (except North Dakota) put a well factor into their depth schedules, that is to say, allowables are not proportionate to acreage. We have also described in several places the relationship of well spacing to economic efficiency and excess capacity. It remains, therefore, to describe briefly how each of our three states handles the problem of determining the size of drilling units and/or production or proration units.

It is essential for any state that has a proration system to formalize the procedure for establishing proration units. This is closely tied to the establishment of drilling units, but the relationship needs to be clarified.

The Oklahoma regulations provide that wells drilled for oil at more than 2,500 feet in depth shall be located not less than 330 feet from any property line and shall be located not less than 600 feet from any producible or drilling well in the same common source of supply.[49] For depths of less than 2,500 feet the minimum distances are 165 feet from any property line and 300 feet from any producible or drilling well. The effect of these minimum-distance regulations is to establish *minimum* spacing of 10 acres for wells in excess of 2,500 feet and of 5 acres for wells of less than 2,500 feet. In addition, the commission may establish well-spacing and drilling units by special orders, and these orders take precedence over the minimum spacing set forth above.

The Oklahoma statutes go into more detail on the establishment of well-spacing and proration units than most states. The basic authority is found in the following passage from the statute:

Whenever the production from any common source of supply of oil or natural gas in this State can be obtained only under conditions constituting waste or drainage not compensated by counter-drainage, then any person having the right to drill into and produce from such common source of supply may, except as otherwise authorized or in this Act provided, take therefrom only such proportion of the oil or natural gas that may be produced therefrom without waste or without such drainage as the productive capacity of the well or wells of any such person considered with the acreage properly assignable to each such well bears to the total productive capacities of the wells in such common source of supply considered with the acreage properly assignable to each well therein.

Establishment of Units. To prevent or to assist in preventing the various types of waste of oil or gas prohibited by statute, or any of said wastes, or to protect or assist in protecting the correlative rights of interested parties, the Commission, upon a proper application and notice given as hereinafter provided, and after a hearing as provided in said notice, shall have the power to establish well spacing and drilling units of specified and approximately uniform size and shape covering any common source of supply, or prospective common source of supply, of oil or gas within the State of Oklahoma.[50]

An amendment to this provision authorizes the commission to approve the drilling of additional wells and to reform units in light of whether oil or gas is found. The only constraint is that the oil or gas pool have relatively uniform spacing.[51]

The law also states that the commission shall hold a hearing to determine the proper size of the drilling unit and shall base its

---

[49] Oklahoma Rules and Regulations, Rule 202.
[50] 52 O.S.A., Sec. 87.1.
[51] Amendment to 52 O.S.A., Sec. 87.1, passed June 3, 1963.

decision on, among other things: (1) The areal extent of the land overlying the pool; (2) the well-spacing plan in use or being contemplated; (3) the depth of expected production; (4) the character of the producing formation; and (5) relevant geological or scientific data. Protection of correlative rights must also be considered. The commission retains the right to increase or decrease size of the spacing units, if the facts warrant such action.

In an amendment passed in 1959 by the Oklahoma Legislature, the statutory provisions for well spacing were altered somewhat and now read as follows:

The Commission shall not establish well spacing units of more than forty (40) acres in size covering common sources of supply of oil the top of which lies less than 5,000 feet below the surface.... The Commission shall not establish well spacing units of more than eighty (80) acres in size covering common sources of supply of oil in the top of which lies less than 9,990 feet and more than 5,000 feet below the surface....[52]

Previously the law had specified a maximum spacing unit size of 40 acres for pools shallower than 9,900 feet. The amendment allows up to 80 acre-spacing for pools 5,000 to 9,990 feet in depth, a much more reasonable provision in the light of the expense involved in drilling wells in this depth range.

Thus, in Oklahoma we find minimum and maximum limits for well spacing set by statute. For wells shallower than 2,500 feet the commission must establish spacing somewhere between 5 acres and 40 acres. For wells deeper than 2,500 feet but less than 5,000 feet, it must establish spacing somewhere between 10 acres and 40 acres. For wells between 5,000 feet and 9,990 feet, the spacing must be between 10 acres and 80 acres. And for wells deeper than 9,990 feet, the spacing minimum is 10 acres, and there is no set maximum. Comments from industry sources indicate that 80-acre maximum spacing on 5,000- to 9,900-foot wells in western Oklahoma has created problems in areas where 160-acre spacing would be more appropriate. To get 160-acre spacing there must be voluntary lease- and royalty-owner agreements and special allowables set by the commission. The commission, under these statutory requirements, has considerable flexibility although not as much as in most other states in which complete discretion in spacing is given to commissions. From the evidence we have been able to find it would appear

---

[52] 52 O.S.A., Sec. 87.1 (c).

that the Oklahoma commission has not fully exploited the maximum-spacing provisions.

Related to the general spacing requirements is another statutory provision that provides a bridge between the well-spacing and proration regulations. This law provides for compulsory pooling of drilling units. It reads, in part, as follows:

When two (2) or more separately owned tracts of land are embraced within an established spacing unit, or where there are undivided interests separately owned, or both such separately owned tracts and undivided interests embraced within such established spacing unit, the owners thereof may validly pool their interests and develop their lands as a unit. Where, however, such owners have not agreed to pool their interests, and where one such separate owner has drilled or proposes to drill a well, on said unit to the common source of supply, the Commission, to avoid the drilling of unnecessary wells, or to protect correlative rights, shall upon a proper application therefor and a hearing thereon, require such owners to pool and develop their lands in the spacing unit as a unit. All orders requiring such pooling shall be made after notice and hearing and shall be upon such terms and conditions as are just and reasonable and will afford to the owner of such tract in the unit the opportunity to recover or receive without unnecessary expense his just and fair share of the oil and gas. The portion of the production allocated to the owner of each tract or interests included in a well spacing unit formed by a pooling order shall, when produced, be considered as if produced by such owner from the separately owned tract or interest by a well drilled thereon. Such pooling order of the Commission shall make definite provisions for the payment of cost of the development and operation, which shall be limited to the actual expenditures required for such purpose not in excess of what are reasonable, including a reasonable charge for supervision. In the event of any dispute relative to such costs, the Commission shall determine the proper costs after due notice to interested parties and a hearing thereon. . . .[53]

This law permits voluntary pooling in drilling units and requires it if it is not voluntary. The allocation of costs and production on the pooled property is left up to the commission, subject to judicial review. This allocation is usually on the basis of the acreage owned by each of the pooling interests. Such a regulation has the obvious benefit of reducing the number of unnecessary wells drilled. It also has the advantage of assuring the interests in the pooled unit of fair treatment with respect to costs and production. "Fair treatment" is a matter of opinion and can be contested in the courts.

[53] 52 O.S.A., Sec. 87.1.

In general the courts are reluctant to upset a commission order of this type without a clear showing of failure to hold an adequate hearing or of arbitrary action on the part of the commission.

Oklahoma also has a law providing for voluntary and compulsory unitization of pools. Under compulsory unitization, if the working interest and royalty interest owners of 63 per cent of the affected property agree, the property will be unitized. This was discussed in an earlier section.[54] Unit operations create some problems in interpool- and intrapool-allowable determination because with such operations the commission may lose its anchor, the depth schedule, for determining well and thus pool allowables. In a unitized field or portion it is not uncommon to find some wells plugged and perhaps others converted to water or gas injection or salt water disposal wells—so it is difficult to devise a per-well allowable system. Thus the intrapool-allocation system breaks down and is replaced by the unit agreement for sharing the pool's production. The commission is left with the problem of determining the pool allowable, since the summing of well allowables is no longer relevant. In the case of unitized secondary-recovery projects this presents no real problem since these projects are usually exempt from market-demand restrictions anyway. For other unit operations—pressure-maintenance projects, for example—the difficulties are quite real. The commission has some obligation to establish "fair" interpool allowables. There is no easy solution to this.

While this discussion is in the context of Oklahoma regulations, other states have these problems. Texas has only voluntary unit operations, while Louisiana has both voluntary and compulsory unit operations. Under any kind of unit operations the commission is faced with problems of setting allowables that are fair to all pools in the state.

The regulations in Louisiana follow the same general pattern that is found in Oklahoma, although there is more flexibility in Louisiana because the statutes do not specify minimum and maximum size drilling units for different depth ranges. The statute, in enumerating the powers of the commissioner, states that he is authorized "to regulate the spacing of wells and to establish drilling units, including temporary or tentative spacing rules and drilling units in new fields." It is further provided that:

---

[54] See Chapter 4.

For the prevention of waste and to avoid the drilling of unnecessary wells, the commissioner shall establish a drilling unit or units for each pool. A drilling unit, as contemplated herein, means the maximum area which may be efficiently and economically drained by one well. This unit shall constitute a developed area as long as a well is located thereon which is capable of producing oil or gas in paying quantities.[55]

The law further states that "the well location for each drilling unit [shall be fixed] so that the producer thereof shall be allowed to produce no more than his just and equitable share of oil and gas in the pool. . . ." And "a producer's just and equitable share . . . is that part of authorized production of the pool . . . which is substantially in the proportion that the quantity of recoverable oil and gas in the developed area of his tract or tracts in the pool bears to the recoverable oil and gas in the total developed area of the pool. . . ."[56] These provisions are rather specific in requiring that the commissioner establish drilling and proration units of a size that can be drained efficiently by one well and in setting well allowables in a pool so that a well's share of the pool allowable is in proportion to that well's share of the recoverable oil in the pool.

In late 1960, Louisiana made a rather radical departure from the conventional method of setting the size of units. The order,[57] in effect, establishes minimum 40-acre spacing for new pools by requiring that a second or confirmation well in a new pool be drilled no closer than 1,320 feet from the discovery well and 600 feet from any property line. However, this is a temporary spacing order, and after two wells have been completed in a new pool, the commissioner will issue no additional drilling permits until a public hearing has been held to establish field rules including, if necessary, spacing regulations. The reasons for this procedure are set forth as follows:

The above requirement is felt necessary since frequently upon completion of a well in a "new pool" other operators owning offsetting acreage to this discovery well make application for the drilling of an offset well. In many instances several applications are filed and should all of the applications be approved at a minimum distance of 1,320 feet from the discovery well then the spacing pattern will have already been established for this "new pool." If it is the intention of the operators to develop this pool on

---

[55] 30 Louisiana Revised Statutes of 1950, Sec. 9B.

[56] *Ibid.*, Sec. 9C and 9D.

[57] Louisiana Department of Conservation, Statewide Order No. 29–H, November 10, 1960.

a 40-acre density, then no harm could be done by the issuance of such permits; but if any operator or operators have intentions of developing this pool on acreage larger than 40 acres then the subsequent permits should not be issued until the hearing as required in Section b hereafter shall have been heard and appropriate rules and regulations including the creation of drilling units for this "new pool" shall have been issued, or unless the proposed well locations comply with the minimum distances from wells and lease lines as prescribed in Finding 6 (b) of this order.[58]

Under Louisiana regulations the initiative is placed with the operators in a field to establish a workable spacing pattern. It cannot be less than 40 acres without commissioner approval.

Louisiana also has a compulsory pooling statute that requires under certain conditions pooling of tracts to form drilling units. It states that:

Where the owners have not agreed to pool their interests, the commissioner shall require them to do so, and to develop their lands as a drilling unit, if he finds it to be necessary to prevent waste or to avoid drilling unnecessary wells.

All orders requiring pooling shall be made after notice and hearing. They shall be upon terms and conditions that are just and reasonable and that will afford the owner of each tract the opportunity to recover or receive his just and equitable share of the oil and gas in the pool without unnecessary expense. They shall prevent or minimize reasonable avoidable drainage from each developed tract which is not equalized by counter drainage.[59]

The system in Louisiana seems to establish minimum 40-acre spacing, with denser spacing a rare exception and requiring a showing by the operators in a specific pool that it is necessary. Wider than 40-acre spacing is encouraged in many instances.

The Texas statutes are much more general in nature than those of either Oklahoma or Louisiana with respect to drilling and proration units. The statute reads, in part, as follows:

The Commission shall make and enforce rules, regulations or orders for the conservation of crude petroleum oil and natural gas and to prevent the waste thereof, including rules, regulations or orders for the following purposes:

1) To prevent the waste, as hereinbefore defined, of crude petroleum oil and natural gas in drilling and producing operations and in the storage, piping and distribution thereof.

---

58 *Ibid.*, Finding 7a.
59 30 Louisiana Revised Statutes of 1950, Sec. 10(A).

2) For the drilling of wells and preserving a record thereof.

3) It shall do all things necessary for the conservation of crude petroleum oil and natural gas and to prevent the waste thereof, and shall make and enforce such rules, regulations or orders as may be necessary to that end.

The Railroad Commission shall also make and enforce rules, regulations and orders in connection with the drilling of exploratory wells and wells for oil or gas or any purpose in connection therewith; the production of oil or gas. . . .[60]

With these general mandates, the commission has passed rules establishing drilling and proration units. Rule 37 establishes statewide spacing. It reads, in part, as follows:

No well for oil or gas shall hereafter be drilled nearer than twelve hundred (1,200) feet to any well completed in or drilling to the same horizon on the same tract or farm, and no well shall be drilled nearer than four hundred sixty-seven (467) feet to any property line, lease line, or subdivision line; provided that the Commission, in order to prevent waste or to prevent the confiscation of property, may grant exceptions to permit drilling within shorter distances than above prescribed when the Commission shall determine that such exceptions are necessary either to prevent waste or to prevent the confiscation of property.

The aforementioned distances are minimum distances to provide standard development on a pattern of one (1) well to each forty (40) acres in areas where proration units have not been established.

In the interest of protecting life and for the purpose of preventing waste and preventing the confiscation of property, the Commission reserves the right in particular oil and gas fields to enter special orders increasing or decreasing the minimum distances provided by this rule.[61]

As the language indicates, this rule sets up 40-acre well-spacing[62] requirements unless the commission rules otherwise, and it requires 40-acre spacing in new pools until a hearing is held to establish specific pool rules.[63] Rule 37 has gained most of its fame for the "exceptions" to the general statewide spacing order that have been granted. Owners of leases on "small tracts" are allowed to drill a well as an exception to Rule 37 if the small tract is part of a subdivision set up before this general spacing order went into effect. The Texas courts have upheld this right to drill.

---

[60] 120 Revised Civil Statutes of Texas, Article 6029.

[61] Recompilations of Rules and Regulations for the Texas Railroad Commission, Oil and Gas Division, 1963, Rule 37, Statewide Spacing.

[62] It is possible to get somewhat smaller than 40-acre spacing using these minimum distances.

[63] A number of salt-dome fields are granted exemption from Rule 37 in the Rules and Regulations.

Because of court decisions that have stripped from small-tract operators the advantages they once enjoyed,[64] the Texas Legislature in 1965 passed a compulsory pooling law that would enable the commission to force small tracts to join to form a drilling unit.[65] The proration system, in recent years, has become a powerful tool with which to reduce small-tract drilling, and it was through this device that compulsory pooling was eventually forced. This will be discussed below.

The Texas commission does have the power to establish proration units in all fields and does so in most cases early in the life of the field. Again there are no specific statutory provisions on how this is to be done—only the general mandate quoted above. Once a pool allowable is established, the commission must allocate in such a way that each producer knows what he can produce. Allocation must also be on a "reasonable basis," and in the *Halbouty* case the Texas Supreme Court has stated that the constitutional requirement of fair shares is to be interpreted as meaning allocation on the basis of "recoverable reserves" underlying each property.[66] In many fields this can be approximately accomplished by uniform well spacing, although there are important exceptions to this. The commission rules and regulations provide for hearings to be held on new fields and for the proration unit and proration formula to be set for each field at that time.[67] In practice, operators frequently apply to the commission for temporary field rules immediately after the discovery well or the confirmation well has been drilled. The temporary spacing established at this time, is, in most cases, adopted in the permanent rules for the field. Once special field rules have been adopted and the size of the proration unit established, Rule 38 provides that:

No well shall be drilled on less acreage than that required for a standard proration unit as established in the applicable rules for any oil or gas field except as hereinafter provided.

.   .   .   .

[64] The so-called Normanna Order created considerable pressure for operators to pool. See 14 Oil and Gas Reporter 885, Rule 3b (1961), *Halbouty v. Railroad Commission*, 357 S.W. 2d 364 (Texas 1962).

[65] Mineral Interest Pooling Act, Vernon's Annotated Civil Statutes, Article 6008c, approved March 4, 1965.

[66] 357 S.W. 2d 364 (Texas 1962).

[67] See Rules and Regulations, *op. cit.*, Rules 39, 40, 42, 43, and 45.

The Commission, in order to prevent waste or to prevent the confiscation of property, may grant exceptions to the density provisions as stated above.[68]

We find, then, that the Texas commission can establish proration units but cannot, in all cases, establish drilling units that conform to the proration units. Until recently, this created anomalous situations for the commission. With the power to establish proration units and the power to assign allowables to these units, it would seem that the commission could assign allowables on a pro-rata acreage basis to wells on tracts smaller than the standard proration unit for a field. However, in 1946, a Texas Court of Civil Appeals in the *Hawkins* case declared that the owner of a small tract not only had the right to drill a well on his property, but also was entitled to an allowable that would result in the recovery of the costs of drilling and producing and return the operator a reasonable profit.[69] This doctrine put pressure on the commission to grant allowables that would result in the recovery of more oil than was originally recoverable under a tract, or, in other words, allowables that resulted in "legalized" drainage of adjoining tracts.

A series of decisions in 1961, 1962, and 1963 by the Texas courts have nullified the *Hawkins* doctrine so that, while a small-tract owner may still have the right to drill on his tract, he is no longer assured of an allowable that will enable him to recover his costs plus a reasonable profit.[70] The courts have insisted that aggrieved parties make their complaint in a reasonable time. The smaller operators and royalty owners who have effectively blocked the passage of a compulsory pooling law in the Texas Legislature found themselves in a position of unequal bargaining power with large-tract owners. This was made clear by the "Patricia rule" of the commission which says, in effect, that no special allowables will be granted to a small-tract owner if he refused a bona fide offer of adjoining tract owners to pool his interests with theirs on a fair and reasonable basis.[71] The results of these events forced support

---

[68] *Ibid.*, Rule 38.

[69] *Railroad Commission v. Humble Oil Co.*, 193 S.W. 2d 824 (Texas 1946).

[70] *Atlantic Refining Co. v. Railroad Commission*, 346 S.W. 2d 801 (Texas 1961); *Halbouty v. Railroad Commission*, 357 S.W. 2d 364 (Texas 1962); *Railroad Commission v. Shell Oil Co.*, 369 S.W. 2d 363 (Texas Civ. Appeals 1963); *Railroad Commission v. Aluminum Co. of America*, 368 S.W. 2d 818 (Texas 1963).

[71] For an application of this rule, see Texas Railroad Commission, Rule 37, Case No. 55, 152 (August 5, 1963).

for a compulsory pooling bill in the Texas Legislature in 1965 by many who had previously opposed such a bill.

The compulsory pooling bill as passed by the Texas Legislature contains what independents consider to be protection of their interests. The language clearly demonstrates their distrust of the major producers. The law states that no one shall be forced to pool if the commission ". . . finds that a fair and reasonable offer to pool voluntarily has not been made. . . ." It also states that:

A pooling agreement, offer to pool, or pooling order shall not be considered fair and reasonable if it provides for an operating agreement containing any of the following provisions:

(1) Preferential right of the operator to purchase mineral interests in the unit;
(2) A call on or option to purchase production from the unit;
(3) Operating charges which include any part of district or central office expense other than reasonable overhead charges;
(4) Prohibition against non-operators questioning the operation of the unit.

. . . .

For the purpose of determining the portions of production owned by the persons owning interests in the pooled unit, such production shall be allocated to the respective tracts within the unit in the proportion that the number of surface acres included within each tract bears to the number of surface acres included in the entire unit. . . . The pooling order . . . shall, as to each owner who elects not to pay his proportionate share of . . . costs in advance, make provision for reimbursement solely out of production, to the parties advancing the costs, of all actual and reasonable . . . costs plus a charge for risk not to exceed 100% of . . . costs.[72]

Under such a law the small-tract owner is protected from loss of production under his lease in that he can force pooling. Also, any pooling offer made to him or by him must be "reasonable." He is assured a pro-rata share of production at least equal to his share of the acreage. If a dry hole is drilled and the small-tract owner has refused to contribute to the costs of drilling, presumably he would not be obligated to pay for any of the cost, since costs can be reimbursed only out of production.

While the problem of "laches," or neglect to take action in a reasonable time, put some restraint on operators who feel that they have been mistreated under existing allocation orders, it is likely

---

[72] Vernon's Annotated Civil Statutes, Article 6008c, approved March 4, 1965.

that the commission could validly change an old allocation order so that future allowables would be on a "fair share" basis.

The picture in Texas can perhaps be summarized by saying that 40-acre or larger proration units have become the general rule in Texas for the future, and that the small-tract problem seems to be a thing of the past in new pools, provided aggrieved parties promptly seek relief from orders that they consider unreasonable. This is a step forward for Texas oil conservation along the long, slow road toward economic efficiency.

Kansas remains the only major producing state that does not have a compulsory pooling statute, barring California, which has no regulatory commission of the type we are discussing.

Chapter 7

Special Topics Related to Proration

1. "MARGINAL WELLS" AND PRORATION

Marginal wells, or so-called "wells of settled production," have been of special concern to the regulatory authorities in most states almost from the beginning of conservation activity.[1] Also, they have been a major source of criticism from economists and others who point to such wells as an example of economic waste created by the existing system of regulations. They thus occupy a critical position in any analysis of the industry conducted from the standpoint of economic efficiency. These wells have been mainly those reduced to pumping after substantial dissipation of the natural driving energies, or new wells in poor structure or along the edges of better structures. "Stripper wells" is sometimes used as a synonymous term, but one has to be careful of definitions as the terms are sometimes used in technical ways. For example, the National Stripper Well Association, for purposes of its statistical survey, defines "stripper wells" as those averaging 10 barrels or less per day, or in fields averaging 10 barrels or less per well per day. It should be noted that the definition goes to actual daily production and not to daily productive capacity, although the two are usually the same for such wells.

Since the general practice of proration began, "marginal wells" in ordinary industry usage have been those to which the rules of

---

[1] Unless specifically noted otherwise, the term "marginal" refers to wells of relatively small and settled production. These may or may not be marginal in an economic sense.

allocated allowables are not applied because of their low producing capacity. Any state can fix as it wishes the maximum amount of producing capacity to be included under this exemption. "Marginal" or "stripper" is, therefore, mainly a regulatory category of freedom from proration, and is quantitative only as defined for this purpose by each state. Such wells are not necessarily "marginal" in any economic sense, that is, a well for which incremental cost-per-barrel equals price-per-barrel. As a matter of fact, it is obvious that many wells classed as marginal or stripper by statute or regulations are extremely remunerative and substantially above the economic margin. A statistical summation of total unrestricted production from low-capacity wells in all states would not be the amount produced by all wells below a stated producing capacity or wells near the stage of abandonment. Moreover, some states do not use market-demand proration and therefore do not define an unrestricted maximum for marginal or stripper wells.

A rough indicator of the relative importance of low-capacity wells in various states is supplied by the *National Stripper Well Survey* of January 1, 1963,[2] using the 10-barrels-per-day test. For the country as a whole, 20.2 per cent of total production is reported as coming from this source, which included 401,031 wells or 68.2 per cent of all producing oil wells. The different states, however, are affected differently in the widest degree. Taking the seven largest producing states, accounting for about 85 per cent of national production, Kansas was reported to get 59 per cent of its production from stripper wells and Oklahoma 58 per cent; Texas stood at 16 per cent and California at 18 per cent; New Mexico was at 10 per cent; and Wyoming and Louisiana were at about 4 per cent. At the extreme limit, Pennsylvania and Illinois were reported as getting over 95 per cent from this source.

Even these figures do not tell the whole story. The stripper category includes much production from secondary-recovery projects, but there is a substantial volume of production in secondary-recovery projects from wells not now included in the stripper category, which would be there (or abandoned) except for the projects. Looking at the matter from a different angle, the existing stripper well fields represent much of the future potential territory for

---

[2] Prepared by the IOCC and the National Stripper Well Association (Oklahoma City, 1963).

secondary-recovery projects. The *National Stripper Well Survey* for 1963 credits such fields with 3.66 billion barrels of primary reserves and 3.31 billion barrels of secondary reserves, or a total of 6.97 billion barrels. Whatever the accuracy or meaning of these figures, the stripper category occupies a position of significance in the total industry picture.

Stripper and marginal wells, however defined, have persistently entered into discussion of regulatory policy and practice since the beginning of conservation regulation. In the resolutions passed at the first meeting of the Oil States Advisory (Governors') Committee in 1931, they were referred to three times. In the first "whereas" appears the "threatened great waste of crude oil reserves by the forced abandonment of approximately 300,000 small wells in producing areas of the United States alone." The other references were as follows:

Fourth: That in all such negotiations and in the dealings of the various State and National authorities with the oil business it be immediately and widely recognized that a fair price for crude oil is essential to its conservation so as to prevent...the abandonment of countless wells which would otherwise produce large quantities of crude oil that could never be recovered if those wells should be abandoned;

. . . .

Eleventh: That in all of such allocations, or between producers as between fields, due regard shall be given to the necessities of the so-called marginal or small producers—those wells which are in danger of abandonment if production or price is curtailed too far—striving in all reasonable manners to preserve and continue the life of all such small wells and fields so that they may produce the amounts of oil consistent with preventing waste therein; ...[3]

Marginal wells are no longer subject to the threat to their survival that was felt to be so great at the early period—the threat of disastrously low and widely fluctuating prices. Nevertheless, they have remained the object of special concern under proration systems, which takes the form of permission to produce to capacity up to some stated maximum. The basis for concern for marginal wells and the rationale given for their preferred treatment has undergone a subtle yet distinct change over the years. Whereas there was originally a concern for the losses of profits of producers

---

[3] *The Compact's Formative Years, 1931–35,* IOCC (Oklahoma City, 1955), pp. 3–6.

and the losses of "economically recoverable" reserves by society as a result of drastically depressed prices in times of glut, the concern today is usually couched in terms of losses of reserves essential to national security and of a weakening of the "small independent" who is said to be the backbone of the domestic producing industry. Invariably the current arguments for protecting marginal wells are in terms of the "physical waste" if such wells are abandoned. We shall want to explore in more detail some of these arguments as well as some of the arguments for "getting rid" of marginal wells. First, however, we must describe the nature of the regulations pertaining to low-capacity wells.

In Oklahoma, for example, the maximum allowed production for nonprorated wells is equal to the lowest current allowable for allocated wells, which has in recent years been around 8 or 9 barrels per day. Exceptions to this may be made by the commission. In Texas, unrestricted production for nonflowing wells is fixed by statute between 10 barrels and 35 barrels per day depending upon well depth. In Kansas, every well has a statutory right to produce up to 25 barrels per day, and proration schedules only apply above that level. In Louisiana, a well is allowed to produce up to its maximum capacity if it is incapable of producing the current allowable at any depth bracket.

Such maxima have no necessary relation to the conditions of survival for such wells. We are unable to find any rationale at all to support the specific regulations on these wells in the various states.[4] In the 1963 *National Stripper Well Survey,* the 401,031 wells producing under 10 barrels per day are shown to average 3.79 barrels, while 67,823 wells in the states of New York and Pennsylvania averaged less than one-half a barrel per day, and the 29,391 wells in West Virginia and Ohio averaged less than one barrel per day. How low production could go at existing prices without requiring abandonment would naturally depend upon particular circumstances, but it is certain that present unallocated maxima for marginal wells are not set by considerations of survival.

The main clue to the way in which marginal-well exemptions are treated is probably to be found in an account of experience in

---

[4] The statutory maximums that define marginal wells in Texas were established in 1935 and have not been changed since. If there was a rationale when the system was initiated, it seems unlikely that it would still apply 30 years later.

Oklahoma in 1930, when statewide proration was just beginning, and allowables were to be based upon well potential officially measured.

It was found that in the flush and semi-flush pools there were 3,095 wells with an aggregate potential of 1,111,000 barrels, and that outside of those areas there were 60,000 older and smaller wells within the state producing approximately 251,000 barrels or an average of 4.18 barrels per well per day; that the fact that the old small wells, commonly referred to as "stripper wells," were so large in number and so scattered throughout the state that their proration would present an insuperable administrative difficulty; and, furthermore, that if prorated, the degree of their curtailment would yield a negligible contribution to the amount of reduction necessary within the state as a whole; and, in addition to this, that to curtail the production of these stripper wells would likely damage them and cause many of them to be abandoned by reason of water-encroachment or otherwise.[5]

Administrative difficulties and negligible benefits may provide sound reasons for excluding large numbers of small wells from the proration system, but there are other reasons for the rules in various states. One reason is pure inertia. Once a system is started, often in haste, it tends to perpetuate itself by creating expectations and vested interests based on it. Moreover, there are far more small wells than big ones, and political pressure by owners has no doubt been exercised to secure favorable treatment.

Because of developing strains in the general proration system, the existing exemptions have been the object of increasing criticism. The source of the pressure is the substantial proportion of exempted production, as for example in Texas where exempted wells account for around 40 per cent of total production. With a rather slowly growing demand factor, all new production has to be accommodated into the other 60 per cent, driving down the percentage rate of use of existing prorated capacity. Apart from the grievance felt by owners of existing prorated wells, an important result is declining incentives for exploration and new development. Assuming for the moment that it is desirable to preserve the exempt wells, the question to be raised is how much fat is there in the exempt categories that might be cut off and fed to the allocated wells. We venture no quantitative opinion on this subject.

---

[5] *Legal History of Conservation of Oil and Gas: A Symposium,* Section of Mineral Law, American Bar Association (Baltimore, 1938), p. 164.

There is no doubt some. To make such a judgment one must know a great deal more about capacities and costs of operating these wells than is now known by the agencies concerned. In addition to the almost complete lack of essential data, there is the likely inadequacy of presently structured commissions to handle any major effort in this direction. This would be particularly true in the market-demand states that have so many thousands of wells. To calculate continually the amount of production that is sufficient to keep each marginal well alive, but no more than this amount, is an administrative impossibility, at least under present organizations.[6] In any case, the problems of adapting the proration system to changing circumstances go far beyond the marginal-well problem and are dealt with elsewhere in this study.

The subject of marginal wells and what should be done about them has from time to time been analyzed by economists, most of whom, it should be added, have a sincere interest in improving the efficiency and health of the industry and the individual companies in the industry. The most recent and detailed study of this sort was made by Professor M. A. Adelman.[7] The general argument runs that there are a large number of high cost, low capacity wells that stay alive solely because of the combined effects of unrestricted production and the maintenance of a relatively high, stable price level. At the same time, production from higher capacity, lower-cost wells is cut back to make room for these marginal wells. Using standard industry cost data from the *Joint Association Survey*,[8] and generally accepted price and well information, Adelman calculates that a closing down of stripper wells in Texas would result in a loss of present capital value to producers of about $1.12 billion while the capital value gain to the nonstripper wells would be about $2.41 billion, which comes to a net gain for Texas producers of about $1.29 billion if stripper wells could be eliminated.[9] For

---

[6] The state of Kansas, however, currently requires annual "potential tests" on all wells. This is a start in determining well capacities.

[7] M. A. Adelman, "Efficiency of Resource Use in Crude Petroleum," *Southern Economic Journal,* Vol. 31 (October, 1965), pp. 101–22.

[8] API, IPAA, MOGA (Mid-Continent Oil and Gas Association), *Joint Association Survey,* Parts 1 and 2.

[9] Adelman, *op. cit.,* see especially Appendix 2. We make no attempt to detail Adelman's calculations. We urge a careful reading of the article for those wishing a full grasp of the analysis.

the entire nation, Adelman roughly estimates a gain of about $3.75 billion in capital values if stripper-well production were transferred to nonstrippers.

More important, in Adelman's opinion, are the losses to be incurred in the future from the drilling of unnecessary stripper wells. An end to this superfluous drilling, he feels, would result in an *annual* saving for the nation of about $2.5 billion. Adelman is careful to point out that his data sources are woefully inadequate and that he has erred on the conservative side in estimating the losses. He is quite aware of the national security issue so often raised in defense of stripper-well production and points out, correctly we feel, that in fields that have overdrilling, abandonment will occur earlier than if the overdrilling had not been done. "Thus, if one well were drilled into a pool and it cost $10 daily to operate, it would last until daily output was worth no more than that amount; but two wells would need $20, and four wells would need $40. Oil which would still be worth producing with one well becomes too expensive to produce with more; and hence overdrilling causes reserves to be abandoned sooner than necessary—obvious physical waste."[10]

Professor Adelman has, in effect, attempted to quantify an aspect of conservation that has long been discussed in only qualitative terms. His critics, and there have been a great many, thus far have answered him primarily with diatribes and with attempts to discredit him as someone who "just doesn't understand the industry."[11] Nowhere has there yet appeared a quantitative rebuttal of his evidence or analysis. From our knowledge of published data sources in the industry, we doubt that any rebuttal can be made other than one using the same statistical series as Adelman used. If this is so,

---

[10] *Ibid.*, p. 122.

[11] An example of this can be found in an address by Frank N. Ikard, President, API, "Petroleum Energy for Today and Tomorrow," before the Houston API Chapter, June 1, 1965. In talking about the critics of conservation regulation, who include others besides Professor Adelman, Mr. Ikard states that "They see prorationing, particularly, as an artificial and harmful device that props up prices and keeps 'uneconomic' units in operation. They see the only overall solution as one that includes capping the strippers, dropping state production controls, junking all or most import restrictions, and allowing operators to produce up to maximum efficient rate whether they have an immediate market or not" (p. 5). We know of no serious critic who has proposed the solutions that are listed by Mr. Ikard.

a rebuttal must rely on the analytical apparatus. We are hopeful that some answer to the Adelman analysis will be made in order to generate a useful dialogue in the troublesome area of marginal wells.

Our own view of Adelman's analysis is that it is helpful in focusing on several problem areas in the conservation system. If this country is interested in such broad policy questions as how much does a healthy domestic oil industry for national-security requirements cost, then it must be interested in determining the whole array of costs, productive capacities, and reserves of existing and prospective wells. In one sense, the Adelman piece can be viewed as a plea for the industry to provide such information to itself and to government policy makers.

The major thrust of the empirical analysis is in two separate, but related directions: (1) the problem of how to handle existing stripper wells, and (2) the problem of how to avoid adding to the number of stripper wells in the future. In the area of existing stripper wells (using this term to cover a number of well categories exempt from market-demand proration), no serious critic in recent years, including Adelman, has recommended an immediate "junking of the system." Yet there are measures which could be taken that are somewhere between doing nothing and blowing up the entire system.

The essence of this problem is that of gradually transforming production from high-cost to low-cost sources. With due advance notice, it would be possible to begin phasing out production from low-yield wells over a period of years. At the same time, much production from stripper fields could be put on a lower-cost basis by combining production into a fraction of the wells and closing the others. We mentioned earlier a proposal of this sort for the East Texas Field. Many of the reserves could be salvaged at lower cost by secondary-recovery operation (discussed in the next section), treating secondary-recovery operations as no different from primary operations, so that they would have to stand on their own economic feet in terms of market-demand restrictions. None of these or similar measures need be radically disruptive. While they would not assure optimum efficiency, they would begin to eliminate the waste and inefficiency that exists in currently operating stripper wells.

The political problem of shifting production from stripper wells to more efficient wells is naturally tremendous. The opposition arises from the damage to vested interests and to local communities in producing areas. So far as direct ownership interests are concerned, it would be possible, as Adelman has shown, to buy them out at a reasonable valuation by a special levy on other producers and still leave the industry in a more profitable position. As for local community interests, the shock could only be softened by phasing the process over an extended period of time.

The second area, that of reducing the drilling of new stripper wells, is somewhat easier to handle. Wide spacing, compulsory pooling, and compulsory unitization early in the life of a field would go part of the way. Advance notice of elimination of exempt status of certain classes of wells would be a great deterrent to the proliferation of wells. Again, none of these types of suggestions would "wreck the system." Unfortunately Adelman has not made any detailed suggestions as to solutions.

The standard answer to critics who suggest that some stripper wells be abandoned is that substantial reserves underlie these fields and to lose them would put a serious dent in the nation's total reserves and thus impair the nation's security. We will note three things in this respect: (1) If an orderly program of phasing out stripper wells could be devised, a substantial quantity of the stripper reserves would still be economically recoverable, particularly if the number of producing wells in each stripper field (and hence their cost of operation) could be eliminated. (2) The abandonment of certain stripper fields and their reserves would be likely to result in an improved financial status for the producing industry and for flush wells in particular. This in turn would increase incentives to exploration and might produce far more reserves at lower cost than the preservation of some of the stripper reserves. Such judgments are conjectural since no one knows the costs and capacities of existing or new wells. (3) If Adelman's figure on the social cost of stripper wells is anywhere near correct, and we make no judgment here on its accuracy except to note that the cost must be substantial, the same amount of liquid-energy supplies felt to be so necessary for national security could quite possibly be developed through a subsidy of a lesser amount to exploration or to the oil shale industry.

The commonly expressed view that the reserves underlying stripper wells are a "precious national heritage" that must be protected by continued operation requires reexamination. As we have suggested, a reasonable policy would not entail the loss of all these reserves. And, to the extent that the wells were phased out, the effect on the remainder of the industry might very well be to stimulate the discovery of new reserves. With due regard to these two factors, the question arises whether the reserves lost are of any particular importance among the multiple sources of energy that will have to be developed to meet the vast energy requirements of the future. The suspicion may be entertained that the "heritage" argument is simply a plausible cover for vested interests.

So much being said, it must still be asked whether it is worthwhile incurring the trouble involved in a positive policy of eliminating stripper wells. Why not let them drag out their history to abandonment, and concentrate on policies designed to prevent the constant renewal of the problem by measures that place the future development of the industry on an efficient basis? On this point we find ourselves ambivalent. It would be difficult to justify the damage to vested interests unless some demonstrable public good were to ensue. Whatever the hypothetical merits of greater economic efficiency, expressed in lower average industry costs, no particular advantage would accrue to the public if the sole result were somewhat higher profits to the remainder of the industry. A public advantage would only appear if higher profits led to more active exploratory effort or if lower costs led to lower prices. On the extent to which these benefits might arise, we have no basis for expressing an opinion. As our earlier analysis has shown, the play of incentives for exploration is bound up with the whole regulatory process, and it is difficult to assess the significance of the single factor of a change of policy with respect to existing stripper wells. That it would work in the direction of greater economic efficiency, a laudable aim, is self-evident. No one but Professor Adelman has attempted to measure the possible economic gain. No one at all has attempted to trace the correlative consequences, in terms of public benefits balanced against the disturbance of vested interests. Discussion of the subject therefore proceeds in an atmosphere of speculative ignorance.

Turning away from policy on existing stripper wells, we turn to the second half of the problem: how to avoid adding to the number of high-cost, inefficient stripper wells in the future. This subject is easier to handle. The desired result would be mainly a secondary consequence of a program of efficient reservoir development through wide well spacing, compulsory pooling, and compulsory unitization early in the life of appropriate reservoirs. Short of the ideal, any progress in these directions will mitigate the problem, which could be further dealt with by eliminating the exempt status of certain classes of wells and phasing out low-yield wells.

2. SECONDARY RECOVERY

Secondary-recovery and pressure-maintenance projects present the regulatory agencies with a difficult dilemma. These projects are designed to augment the natural drive of reservoirs and thus to raise (1) recoverable reserves and (2) producing capacity of these reservoirs. We have already noted that one of the major problems facing the industry is that of finding a way to increase its reserves without at the same time increasing producing capacity. A secondary-recovery project has great appeal because it can "prevent physical waste" by recovering a larger percentage of the oil in place. But, concurrently, these projects add to a state's producing capacity, which means smaller allowables on prorated wells. This is true even if the secondary-recovery project is prorated also. If the project is exempt from market-demand restrictions, then the cutback on prorated wells is even greater.

Thus some people view prospective secondary-recovery and pressure-maintenance projects as attractive sources of reserves that should be developed. Other people view them as factors that further undermine the regulatory system and that create economic hardship for parts of the producing industry, and reduce the incentives to exploratory effort. We shall review these two aspects briefly to point up this dilemma, without electing to sit on either horn of it.

The methods used in reservoir development in the past left a large major fraction of oil in place unrecovered, and, though improved methods have greatly increased the recovery ratios in many newer fields, vast potential reserves underlie developed fields awaiting the application of secondary-recovery methods. How great the reserves will eventually be will depend in part upon recovery technology, in part upon other economic factors of costs and prices, and in part on regulation. There are at present large reserves—estimated in an IOCC report at over 16 billion barrels—that could be profitably produced with present technology and at present costs and prices, and if the heavy burden of excess capacity were not present. But the rate of development is limited by the alternative possibilities offered by alternative lines of investment in exploration and primary development. In part, the relative attractiveness is affected by conservation rules, since both costs and revenues are partly a function of such proration and well-spacing regulations.

It is conventionally stated, with no known factual basis, that early in the century the rate of recovery by the primary methods then used amounted to 15 to 20 per cent of oil originally in place. As estimated by Paul D. Torrey in an IOCC report, a similar figure for 1962 would be about 28.5 per cent.[12] This 1962 figure is arrived at by adding primary (proved) reserves (31.4 billion barrels) to cumulative past production (67.7 billion barrels) and dividing through by estimated original oil in place in presently known fields (346.2 billion barrels). Mr. Torrey then estimates an additional 16.3 billion barrels of secondary reserves recoverable by conventional fluid injection methods under present economic conditions. Adding these into the numerator of the fraction gives a percentage recovery of 33.3 per cent. Torrey then supplies an estimate of 40.2 billion barrels of reserves physically recoverable by known secondary-recovery methods, but not economically feasible under present conditions. If these are added in as hypothetically available at some future time, the recovery percentage would rise to 45 per cent.

In passing, it should be noted that the calculation of the present rate of recovery involves a serious underestimate. The estimates of proved reserves as of any given time tend to double more or less *for the same fields* before the limits of ultimate recovery are reached through "extensions" and "revisions" in the estimates.[13] If these anticipated additions were thrown into the numerator of the Torrey 33.3 calculation, the percentage would be increased to 40. This in turn would be an overestimate since the later extensions and revisions will include some reserves due to secondary-recovery projects, thereby double-counting some of Mr. Torrey's figures.

---

[12] Paul D. Torrey, *Evaluation of United States Oil Resources,* IOCC (Oklahoma City, 1962).

[13] *Background Material for the Review of the International Atomic Policies and Programs of the United States,* report to the Joint Committee on Atomic Energy, Vol. 4, Joint Committee Print, 86th Cong., 2d Sess. (Washington: U.S. Government Printing Office, 1960), p. 1530.

No claim of accuracy can be made for such calculations, and this is especially true of the denominators of the fractions, original oil in place. Though subject to question as to accuracy, the IOCC figures do, nonetheless, serve a purpose in providing rough orders of magnitude for thinking about potential additions to the supply through improved recovery methods and stronger economic incentives. In the case of fully developed reservoirs, the additions will mainly be through secondary recovery. For newer reservoirs, they may arise from more efficient methods of reservoir development.

Mr. Torrey estimates 1960 production from "fluid injection projects" at 732.9 million barrels, or 29.6 per cent of total production.[14] But this includes production from pressure-maintenance projects not properly classified as secondary recovery. For the year 1962, the *National Stripper Well Survey*[15] attributes 591.9 million barrels to 406,102 stripper wells (defined as 10 barrels per day or less), and identifies the portion of this from secondary recovery in some states, but not including Texas or Louisiana.[16] Even if the identification were complete, it would give no clue to total secondary production, since much of it lies outside the stripper category, i.e., wells on secondary recovery producing more than 10 barrels daily. The report on *Unitized Oilfield Conservation Projects*[17] is equally partial. It purports to show "the application of unitization on secondary recovery and pressure maintenance operations" and attributes production of 400.5 million barrels to such projects in 1961. The secondary-recovery portion is not separately identified but, even if it were, it would be partial, since the part outside the "unitized" category is not identified. Mr. Torrey lists 5,734 fluid-injection projects. *The Unitized Oilfield Conservation Projects* lists 1,550 projects, which should in principal be part of Torrey's list.

No one knows how many wells are engaged in secondary recovery, or how many of them are in multiple-well, more or less unitized projects, or how much oil they produce in various states. This is not true of some individual states. Louisiana, for example, publishes

---

[14] The Department of the Interior estimates production by secondary recovery at 29.1 per cent of total production in 1961. See *An Appraisal of the Petroleum Industry of the United States* (Washington: U.S. Government Printing Office, 1965), Table 25.

[15] *Op. cit.*

[16] Comparable data not available for later years.

[17] IOCC (Oklahoma City, 1962).

an annual report on secondary-recovery projects and production. In 1964, 460 projects produced 64.4 million barrels, or 12.1 per cent of total production.[18]

Sparse as they are, the figures available do suggest the line of thought that, among the thousands of older fields, there is a vast untapped field for investment in secondary recovery if and when the economic incentives become attractive. The incentives to put money into secondary recovery vary greatly, depending on the special circumstances surrounding each potential secondary-recovery situation. But certain things of a more general nature can be said about the great producing areas like Texas, Oklahoma, and Louisiana, where secondary recovery has to compete for investment funds against exploration, development, or other uses.

It is interesting to ponder the question (implicit in Torrey's figure of 16 billion barrels of reserves which could be produced with present secondary-recovery technology and at present costs and prices) of why more secondary-recovery projects are not undertaken. This is not to imply that there has not been rapid growth in this area in the past or that this growth will diminish. Informed sources in the industry indicate that combined production from secondary-recovery and pressure-maintenance projects will soon be greater than primary production.[19] Certainly there have been some major breakthroughs in secondary-recovery technology in recent years. Several possible answers to the question seem plausible.

One answer may be that while some secondary-recovery projects may in fact be profitable (incremental cost-per-barrel is less than price), they may, in anticipated results, be less profitable than alternate uses of investment funds. Obviously, investment funds are not unlimited. As we indicated in Chapter 1, the investment process is one in which net income from the investment is maximized over time. Petrochemicals or exploration may possibly be expected to yield higher returns than secondary-recovery operations.

---

[18] Louisiana Department of Conservation, *Secondary Recovery and Pressure Maintenance Operations in Louisiana* (1964).

[19] See the testimony of various witnesses before the Texas Railroad Commission, "Statewide Hearing Pertaining to Water Flood, Pressure Maintenance and MER's," Docket No. 20–43, 172, June 16 and 17, 1960. See also U.S. Department of the Interior, *An Appraisal of the Petroleum Industry of the United States, op. cit.*, pp. 14–15 and Table 28, where secondary recovery is projected at 42 per cent of total production in 1975.

Another answer sometimes given is that companies tend to push exploration for new reserves to assure a continued supply of crude to support integrated company operations. This may have some validity if it implies merely a postponement of the initiation of secondary-recovery operations, since a company could view its undeveloped secondary reserves as assured and capable of development some time in the future. However, reserves are reserves, be they primary or secondary. In a sense, a company is interested in the present and future costs of production. It is not particularly interested in whether this production is primary or secondary except as these designations create some special cost or revenue circumstances, which they may well do under present regulations. We shall return to this point in a moment.

The difficulties of getting property interests to agree is perhaps the major reason why there are many secondary-recovery projects that seem to be economical, but are not developed. There apparently are a large number of pools that could be subjected to secondary-recovery methods with success *if* the most efficient plan for development could be used. This is a very large "if," however, and for many fields the use of the most efficient method would require unit operations. Clearly, the location of producing and injection wells is dictated by the nature and shape of the reservoir. If, as is often the case, the location of these wells must be altered by surface property lines to avoid drainage and protect correlative rights, the profitability of such projects may dwindle or even vanish. There are numerous examples of secondary-recovery projects that cover only part of a field. In many such instances the costs are higher and/or the recovery less than if the field had been developed as a unit. Thus, we find ourselves back into a discussion of unit operations.

All this suggests that while Mr. Torrey estimates that there are 16.3 billion barrels of secondary reserves recoverable by conventional fluid-injection methods under present economic conditions, he may implicitly assume the application of the most efficient technique in each field. It is doubtful that he considered the economic barriers posed by property owners in these fields who may not be able to agree on the shares of costs and revenues in potential secondary-recovery projects. He may also have neglected the question of alternative uses and profitableness of investment funds.

In the absence of firm evidence, we hesitate to assert that sec-

ondary recovery is being relatively neglected in favor of investment in discovery and development. The contrary is asserted by
E. Morris Seydell in the text accompanying the *National Stripper
Well Survey*.[20] Having stated that secondary-recovery projects are
largely based on stripper or marginal wells, he goes on to say: "The
marginal well today stands as the backbone and foundation of the
domestic producing industry. We have seen the emphasis change
from an industry exploring and looking for new reserves to one
doing research and improving methods of recovering known reserves. For the period of the short run, little damage will be done
by this change in emphasis, but for the period of the long run, it
will spell disaster for the domestic reserve picture" (p. 13). This
interpretation appears to us to muddle up and oversimplify a
very complex problem with respect to future availability of crude
oil from old reservoirs and from newly discovered ones—both of
which are of importance.

Turning the coin over, it can be argued that we are not at all
negligent of developing secondary reserves; on the contrary, we are
in fact developing too much of them. The industry is faced with
producing overcapacity, yet we find that new secondary-recovery
and pressure-maintenance projects are being initiated at a rapid
rate. This can only add to the already excessive capacity and result
in greater restrictions on prorated wells in market-demand states.
Does it make sense to restrict production of relatively low incremental-cost primary production in order to make room for relatively
high incremental-cost secondary production? Clearly the economics
of these situations governs. Part of the boom in secondary-recovery
operations, and particularly those which are the capacity water-
flood type, is caused by the total or partial exemption that their
production enjoys under present conservation regulations. There
is a clear incentive to go to a water flood if the pool is then exempt
from market-demand restrictions. The question has been raised on
numerous occasions whether or not secondary-recovery projects
should be subject to market-demand factors. A hearing before the
Texas Railroad Commission in 1960 was called specifically to explore this question. Despite much evidence from several large companies that the preferential treatment could be eliminated for
these pools, the commission was undecided and promulgated no

---

[20] *Op. cit.*

statewide rule. It has, apparently, given a closer look since 1960 to applications for secondary-recovery projects and been more reluctant to exempt these entirely.[21]

No one knows how much producing capacity exists in secondary-recovery and pressure-maintenance projects. There is no clear picture of the economics of these situations. From our discussions with industry representatives, it is obvious that a considerable amount of restriction could be imposed on many of these projects and still have them profitable. If this were done, it would be possible to expand production in wells that have market-demand restrictions.

The fact that old reservoirs, inefficiently developed in the first place, are not being redeveloped at a very rapid rate does not seem to us a matter of any importance. In discussing so-called marginal wells at an earlier point, we expressed the view that some of the potential reserves underlying them may be regarded as a valuable social asset. But there is no urgency about developing them in the face of the existing overcapacity in the industry. So long as they remain accessible, they can be developed or not, as economic circumstances warrant. We include in this generalization the realization that for any given field the costs of initiating a secondary-recovery project may be greater tomorrow than today because of the physical deterioration of wells and equipment in the field.

We must, however, note one factor which, if not remedied, might and apparently does seriously deter secondary-recovery operations as and when they would otherwise be economically attractive. This is the absence in many states of compulsory unitization laws. In the absence of such laws the costs of redevelopment are likely to be greatly magnified—often to the point of preventing an effective and economically feasible program.

A Statistical Note

No direct comparison can be made among the three studies cited in this section. However, using the data on Kansas, the following cross-references may be of some interest:

The study, *Unitized Oilfield Conservation Projects,* credits Kansas

---

21 In 1958–60, 354 secondary-recovery projects were initiated in Texas, while in 1960–62, the number declined to 333. *Oil and Gas Journal,* November 18, 1963, p. 136.

with 90 unitized projects producing 3.6 million barrels of crude oil in 1961, or 3.2 per cent of total production. This comes to an average of 40,000 barrels, or 110 barrels per day, per project. The secondary-recovery portion is not isolated.

Paul D. Torrey's *Evaluation of United States Oil Resources* credits Kansas with 833 fluid-injection projects producing 19.3 million barrels in 1960, or 17.1 per cent of total production. This comes to 23,000 barrels, or 63 barrels per day, per project. The secondary-recovery portion is not isolated. There is no necessary conflict with the *Unitized Oilfield* data, which should in principle be a subcategory of Torrey's study.

The *National Stripper Well Survey* credits Kansas with 40,951 stripper wells in 1961, producing 73.6 million barrels, or 65.6 per cent of total production. Of the stripper production, 19.1 million barrels, or 26 per cent, is credited to secondary-recovery projects. This is stated to be 95 per cent of all secondary recovery, bringing the total to 20 million barrels, or slightly larger than Torrey's total for "fluid-injection" projects. The statistical basis of the study is made subject to question when we see that the number of stripper wells (40,951) is larger than the total number of producing wells in Kansas reported by the U.S. Bureau of Mines (40,933).

3. A NOTE ON "MER"

The chaotic conditions of overdrilling and overproduction in the newly discovered fields during the early 1930's forced upon the oil industry an important rule of reservoir behavior—namely, that the *rate* at which a reservoir is produced has an important effect on the proportion of oil originally in place that is ultimately recovered. Put in other words, it became apparent that rapid and indiscriminate production often "wasted" reservoir energy, caused irregular, non-uniform migration of fluids and a by-passing of large deposits, and resulted in "premature" abandonment of the field and reduced ultimate recovery.

There soon evolved an imprecise concept of a maximum rate of production that could be allowed and that would come close to achieving maximum ultimate recovery. The implementation of this concept was usually in terms of prohibiting the inefficient use of natural-reservoir energies, i.e., the setting of maximum allowed gas-oil ratios or water-oil ratios. As early as 1938, a concept of this sort was being used in industry literature.[22] With the accumulation of reservoir information and the technological know-how of reservoir management, the concept of a maximum production rate became sharper. A progress report prepared by a special committee of the API placed great emphasis on producing rates.[23] The report states that:

The prevention of waste is accomplished by the application of the most efficient production methods known and economically available. It will be shown in the following section that the control of production rate is essential to any effective method for improving production efficiency and, therefore, production rate is perhaps the most important single factor which must be regulated. The entire structure of the conservation program rests upon the ability to control properly the rate of production.[24]

Later in this report there was a more detailed discussion of restriction of production to prevent waste:

Production rate is the most important controllable factor in the prevention of waste, and no pool should be allowed to produce at so high a rate as to cause waste. It has been argued that there is an optimum production rate at which each field will produce a maximum amount of oil. It is doubtful that this is true. In any case, engineering skill has not yet developed sufficiently to permit the determination of the optimum rate with the accuracy required for allocation purposes. It is true that a rate can be determined, for most fields, at which waste would occur due to an excessive production rate; but there is a wide range of rates in which efficient operation is possible, and within this range it is not possible to select the exact point of maximum recovery, even if such a point exists.[25]

This statement is evidence of the state of knowledge about reservoir mechanics and correctly points out that there probably is not a single rate of production for a reservoir which, without question, yields maximum recovery. Rather there is more likely a *range* of production rates that allow for efficient operation and maximum recovery. At this point in time, 1942, the industry was close to the

---

[22] See Edgar Kraus, " 'MER'—A History," reprint of a paper presented at the annual meeting of the American Petroleum Institute, Chicago, March 11, 1947, for a discussion of the evolution of this concept.

[23] Special Study Committee on Well-Spacing and Allocation of Production, *Standards of Allocation of Oil Production within Pools and among Pools* (Dallas: American Petroleum Institute, 1942).

[24] *Ibid.*, pp. 17–18.

[25] *Ibid.*, p. 64.

current concept of a maximum rate of efficient production (or maximum efficient rate of production, MER). The only additional step necessary is to indicate that the range has a limit beyond which ultimate recovery is not maximized.

With the advent of World War II, the MER concept took on national defense significance. The task of the oil industry was to supply immediate wartime needs with a minimum expenditure of materials and manpower, and at the same time maintain reserves and productive capacity at the highest possible levels in the event that the war was prolonged. At the beginning of the war there was excess crude-oil-producing capacity, although not always located most advantageously. Curtailment of drilling, the natural decline of reservoir capabilities, and the increased demand eliminated the excess capacity in production. During these years a critical question was what the maximum efficient rates of production were for the various fields in the country. Such information was essential for proper management of oil resources; without it there was no way to know how to get the greatest possible output without reducing ultimate recovery. Industry committees were asked by the Petroleum Administration for War (PAW) to estimate MER's on a field basis. In the case of the estimates made for Texas, Kraus reports that, "How the Ivy committee arrived at its estimate of maximum efficient rate is 'restricted' information." But because of the technical skills and extensive knowledge of the committee members, "Their combined knowledge probably obviated the need of a rigorous definition."[26] By mid-1944, actual production in the United States exceeded "maximum efficient productive capacity."[27] Fields in the Midwest undoubtedly exceeded their MER's before this date, since they were strategically located to supply war needs.

The historians of the PAW define MER in the following way: "The 'maximum efficient rate' is defined as the highest sustainable rate at which a field can be produced for a designated period without appreciable loss in ultimate oil recovery."[28]

Such a statement is clearly not a working definition, but rather a generalized statement intended to give an impression. It does,

[26] Kraus, *op. cit.*, p. 2.

[27] J. W. Frey and H. C. Ide, *A History of the Petroleum Administration for War, 1941–1945* (Washington: U.S. Government Printing Office, 1946), Appendix 12, Table 16.

[28] *Ibid.*, p. 176, n. 7.

however, pinpoint some key factors that must be resolved to arrive at a working definition. Clearly, the MER is a function of many technical features of the reservoir and thus will vary from reservoir to reservoir. Also, it must have a time dimension that takes into account current oil foregone versus ultimate oil foregone.

One of the best statements of the MER concept is found in *Petroleum Conservation,* edited by Stuart E. Buckley.

The requirements for efficient recovery of the oil from a reservoir are not taken care of by chance; they may be fulfilled only through careful and deliberate action by the producers. Experience has shown that one of the most essential factors in meeting these requirements is control of the rate of production. Excessive rates of withdrawal lead to rapid decline of reservoir pressure, to release of dissolved gas, to irregularity of the boundary between invaded and non-invaded sections of the reservoir, to dissipation of gas and water, to trapping and by-passing of oil, and, in extreme cases, to complete loss of demarcation between the invaded and non-invaded portions of the reservoir, with dominance of the entire recovery by inefficient dissolved-gas drive. Each of these effects of excessive withdrawal rates reduces the ultimate recovery of oil.

Operation of a reservoir at an infinitesimally low rate of oil production would not in itself assure efficient recovery unless other necessary conditions were met. However, if all other conditions are fulfilled, the ultimate oil recovery from most pools is directly dependent on the rate of production. The nature of this dependence is such that for each reservoir there is for the chosen dominant mechanism a maximum rate of production that will permit reasonable fulfillment of the basic requirements for efficient recovery. Rates lower than such maximum may permit still higher ultimate oil recovery, but once the rate is sufficiently low to permit the basic requirements to be met, the incremental ultimate recovery obtainable through further reduction of the rate of production may be insufficient to warrant the additional deferment of a return and the additional operating expenses that would result from a prolongation of the operation. A rate of production so low as to yield no return would obviously be uneconomic, of no advantage to the operators, and of no ultimate benefit. However, increase of the rate of production beyond the maximum commensurate with efficient recovery will usually lead to rapidly increasing loss of ultimate recovery. From these considerations there has developed the concept of the *maximum efficient rate* of production, often referred to as the M.E.R. For each particular reservoir it is the rate which if exceeded would lead to avoidable underground waste through loss of ultimate oil recovery.

The M.E.R. is not an invariant characteristic of a reservoir, but is dependent on the recovery mechanism employed as well as on the physical nature of the reservoir, its surroundings, and its contained fluids. For the same reservoir it will be different for one recovery mechanism than for another, and for the same mechanism the M.E.R. "may vary with the degree of depletion. The M.E.R. is determinable through engineering study provided adequate geologic and operating information on the reservoir is available.[29]

Here we have spelled out the engineering considerations relating to reservoir characteristics and performance, field development including well location, the recovery mechanism, production methods, degree of depletion, and the like. In addition, there is included here an economic concept, since ". . . the incremental ultimate recovery obtainable through further reduction of the rate of production may be insufficient to warrant the additional deferment of a return and the additional operating expenses that would result from a prolongation of the operation. A rate of production so low as to yield no return would obviously be uneconomic. . . ." Thus prices of the oil produced currently and in the future, the amount of capital invested in producing facilities, the expected reasonable return on this capital, the discount rate for computing the present value of future income streams, and the costs of operating the facilities become variables in determining the MER.

We have discussed the MER concept as presented in the Buckley volume with industry and regulatory representatives and found that, as the term is generally used, it has none of the economic variables we have noted; or, at least if they are included, they are never made explicit. A casual sampling of Texas Railroad Commission hearings to determine MER's for specific pools reveals that technical conditions are the only topics discussed. Further confirmation of this conclusion is found in the 1964 IOCC Conservation Study which discusses the MER concept in strictly engineering terms.[30] Some economists working in the industry, however, persist in including economic variables in determining optimum rates of production, and they are apt to couch their discussions in terms of MER's. It would perhaps be helpful if those using these concepts could distinguish between the "maximum efficient rate of production" as an engineering concept and a "maximum efficient economic rate of production," or an "optimum economic rate" (OER) as an economic

[29] *Petroleum Conservation*, American Institute of Mining and Metallurgical Engineers (New York, 1951), pp. 151–52.

[30] IOCC Governors' Special Study Committee, *A Study of Conservation of Oil and Gas in the United States* (Oklahoma City, 1964), pp. 59–65.

concept. The OER would include the MER concept as well as the relevant economic variables. It seems to us that the OER (MER as discussed by Buckley) falls precisely within the concept we discussed in Chapter 1 as the economists' view of conservation. Our discussion will touch on both the OER and the MER, since both seem relevant.

It is obvious that the determination of a pool OER in a truly meaningful sense is exceedingly difficult to make and requires numerous assumptions about technical and economic variables in the future. This discussion also raises the uncomfortable problem of what is meant by wishing to produce in such a way that ultimate recovery from a pool may be optimized. The OER is not the rate of production that maximizes the number of physical barrels recovered. The above quote from Buckley indicates this is not what is sought. This leaves us with a concept of a rate of production that maximizes the discounted net flow of revenue from a pool, a concept heavily dependent not only upon the physical reservoir characteristics and recovery practices, but also upon present and future costs, present and future prices, and present and future interest rates. If all these problems could be overcome, we would still be faced with the problem of whose discounted net flow of revenue is supposed to be maximized. Is it the revenue of the ownership interests taken collectively in a pool, or is it some "social revenue" or "social benefits" concept? Either choice presents difficulties. Further discussion in this direction at this point can add little to our current discussion of the economic aspects of proration, but enough has been said to indicate that a great deal more thought needs to be given these concepts.[31]

Let us now reverse our field somewhat and speak of the MER concept as it is used in the industry, i.e., as some ascertainable rate of production beyond which there will be a reduction in ultimate recovery from a reservoir. In some theoretical sense such a concept has much merit in oil conservation. It would be quite sensible to set up a proration system based upon MER's so that, if demand exceeds supply, pools would not be overproduced and thus damaged. Also, there would be a definite rationale, in a situation in

---

[31] For additional discussion on MER, see J. W. McKie and S. L. McDonald, "Petroleum Conservation in Theory and Practice," *Quarterly Journal of Economics*, Vol. 76 (February, 1962), pp. 98–121.

which supply (at MER) exceeds demand, in restricting production according to relative MER's for the various pools in the state. If production is to be restricted 50 per cent, then the pool having a daily MER of 1,000 barrels would be cut to 500 barrels and the pool having a daily MER of 2,000 barrels would be cut to 1,000 barrels. From the standpoint of equity among producers in different pools, it is open to argument whether such an allocation system is *the right* one. It has been argued by some that the giant fields would get the lion's share of prorated production. On the other hand, such a system does have the strength of a logical basis, something that the present system of depth-factor allocation does not have.

Under present conditions of overcapacity, the MER concept has fallen into disuse in Texas, the only market-demand state that attempts to use the MER as a conservation tool. This, however, has not always been the case. The history of the Texas yardsticks gives some interesting insights into the factors which have at times dictated the use of MER's as a basis for interpool allocations and at other times the use of depth-acreage schedules.

Prior to World War II, Texas had no specific yardstick or formula to determine shares by pools.[32] With the advent of the war, the 1941 formula was devised that provided a yardstick for each of five types of crude.[33] In 1942, in order to comply with PAW requests for expanded production of crudes that yielded high octane gasoline, toluene, and other critical products, the Railroad Commission established the 1942 Yardstick which combined the five old yardsticks and was used to determine allowables for new fields until MER hearings could be held. Once an MER had been determined for a field, this became the basis for allocation. In July, 1947, the 1947 Yardstick was established which generally increased allowables under the depth-factor schedule. During the late war years and immediate postwar years there was virtually no market-demand restriction, and pools were produced at MER or according to the Yardstick if MER had not been determined. In 1948, for example, there were no shutdown days in Texas. In the fall of 1949, the demand for Texas crude fell off, and by February, 1950, production was reduced to fifteen days. At that time the commission held hear-

---

[32] The following factual material was taken largely from the testimony of Jack K. Baumel, testifying for the Socony Mobil Oil Company before the Texas Railroad Commission, Docket No. 20–43, 172, June 16 and 17, 1960, pp. 286–99.
[33] Miranda type, low octane, medium octane, high octane, and low cold test.

ings to study the problem of high-MER fields that some people claimed were receiving a disproportionate share of the restricted production. The commission, after this hearing, reduced all fields in the state to the maximum provided in the 1947 Yardstick. Thus, fields with MER's higher than the schedule allowable for their particular brackets were given the yardstick-schedule allowable and then restricted for market-demand purposes from this base, rather than from their MER's.

It is interesting to note that three alternative proposals were made at the 1950 hearings. One was to use the old 1942 Yardstick; the second was to use the 1947 Yardstick; and the third ". . . was to give each well in the State of Texas the marginal allowable, and that marginal allowable was then subtracted from the market demand and what difference existed between the marginal for each well in the State and the market demand was then allocated to a fraction of the MER for the State."[34] The 1947 Yardstick was chosen, and producing days in Texas rose from fifteen in February to seventeen in March.[35] There were 134 fields that had MER's higher than the Yardstick, and the cutbacks in those fields permitted the increase in producing days. The next month, April, 1950, the demand rose, and the commission allowed the 134 MER fields to produce on the basis of their yardstick plus 45 per cent of the difference between their yardstick and their MER. With the onset of the Korean War and the rapid rise in civilian consumption of oil, demand rose so that by August, 1950, the commission reverted back to 100 per cent MER's and allowed twenty-one producing days. Since 1950 the commission has allowed 100 per cent MER as the basis for allocation to these *old,* high-MER fields. In general, *new* fields coming off discovery allowable were given a ceiling of the 1947 Yardstick, although there have been a few exceptions to this. In 1960, one witness before the commission testified that 2.2 per cent of the fields in Texas (153 out of 6,906 fields) had approved MER's higher than the Yardstick, and they accounted for 16 per cent of the schedule allowable. Apparently there has been no general review of Texas MER's since 1943, and, since 1956, there have been very few specific pool MER's reviewed.

This brief historical account suggests a reason for switching from

---

34 Baumel testimony, *op. cit.,* p. 291.
35 The East Texas Field was raised from 13 to 17 days.

the use of MER's as a basis for allocation, when market-demand restrictions were relatively slight, to the Yardstick, when restrictions became tight. If market-demand restrictions are severe, allocations under MER would fall heavily to the rich wells, raising questions of "equity" to the producers of poorer wells. One way to meet this claim is to reduce the production from the profitable high-MER fields and shift part of their allowables to other wells. (Another way would be to restrict production from some exempt wells.) Thus, we find again a sharp clash between efficiency and equity. On equity grounds there may be some arguments for such action. This conflict, however, gets us back to the entire structure of the allocation system.

Prorated wells are currently producing at about 33 per cent of their allowables as determined by the depth-factor schedule. It is generally conceded that many pools could not produce at 100 per cent, so that the capacity or MER of these pools lies somewhere between 33 and 100 percent of the schedule allowable. There are other pools in which the MER exceeds 100 per cent of the schedule allowable. In a few pools the MER rather than the Yardstick is used to determine the allowable. The feeling of the Texas regulatory authorities seems to be that, as long as the market-demand restriction must be kept so severe, there is little need to worry about either MER's or 100 per cent schedule allowables. Since the last general MER hearings in Texas occurred in 1956, for most pools in Texas no one knows the MER's, particularly since a pool MER changes over time as the oil is produced.

Oklahoma and Louisiana make no effort to use an MER concept in their regulations except to avoid setting allowables that exceed "productive capacity" or that would cause "underground waste." In these states, as in Texas, there are many pools that could not sustain production at 100 per cent of their schedule allowables. This fact raises another question about the appropriateness of depth-factor schedules in achieving the waste-avoidance goal. It is evident that the existing schedules are workable today primarily because of the large degree of excess capacity. If conditions arose that required a substantial lessening of restrictions, the proration systems would be beset by more severe problems than now face them. Allowables would be likely to be raised with little consideration of whether or not the new higher production rates were excessive in terms of efficient reservoir management. This, of course, occurs at current producing rates in some fields, but it would be magnified any time allowables were raised.

The only state that currently pays explicit attention to the MER concept is California, but there it is not used as a proration guide or even by a state agency. Production in California is not subject to commission regulation. However, under the aegis of the Conservation Committee of the California Oil Producers, pool engineering committees make MER recommendations and pool operators committees deal with problems of intrapool distribution of the recommended rate of production. An engineering board of the statewide organization reviews these findings and makes recommendations based upon them. These are circulated to members of the industry and are submitted to the California Oil and Gas Supervisor, a state official, who may concur or dissent and make alternative recommendations. None of the findings are legally binding upon operators and apparently are often ignored, although proof of this is difficult to come by.

The relevance of the MER concept to the California situation arises from the fact that the West Coast is an area where market demand is in excess of the regional supply. The wartime MER principles have been continued on a voluntary basis as a device for avoiding excessively rapid depletion of underground supplies to the detriment of ultimate recovery. No one knows how successful they have been.

4. CORRELATIVE RIGHTS AND OTHER ASPECTS OF PROPERTY RIGHTS

The model conservation statute prepared by the IOCC states the purpose of regulation as follows: "The prevention of Waste of Oil and Gas and the Protection of Correlative Rights are declared to be in the public interest. The purpose of this Act is to prevent such Waste and to Protect Correlative Rights." A similar statement appears in some state statutes but even where it does not appear, it necessarily comes in the side door of regulatory action. "Correlative rights" means in effect some definition of rights of access to opportunities of economic gain arising from the presence of oil or gas in a tract of privately owned land. It should be noted at the outset that the law as it relates to title of oil in place is not uniform among the oil- and gas-producing states.[36] However the rules differ in detail, they confer the right on an owner of a tract of land to drill and produce, and to secure title to the oil produced. Case law in each state has played a major role in shaping the specific rules and doctrines governing the rights to oil property, but a review of this case law goes beyond the scope of this study.

Since this study is concerned primarily with economic matters, we shall not enter into the niceties of definition of legal rights. Nevertheless, we must give attention to correlative rights since they may enter deeply into some of the questions of economic efficiency with which we are chiefly concerned.

Allocation to Wells within Pools

As we saw in the earlier description of the proration system, the rights of access are controlled by well-spacing rules and the rights to produce by depth-acreage and proration formulas. These to a large degree define property rights. But state rules commonly specify the protection of property, subject to valid waste-prevention measures, often using the phrase to "protect" or "adjust" correlative rights. Since the several states have very different regulations, there cannot be said to be any uniform operative definition of correlative rights. Against this diversity of practice, we may usefully set the treatment of the concept in the IOCC model statute.

The IOCC model statute approved in 1959 contains the following in the definition section: Rule 1.1.14 " 'Protect Correlative Rights' means that the action or regulation by the Commission should afford a reasonable opportunity to each Person entitled thereto to recover or receive the Oil or Gas in his tract or tracts or the equivalent thereof, without being required to drill unnecessary wells or to incur other unnecessary expense to recover or receive such Oil or Gas or its equivalent." This definition of "protect correlative rights" at once links property rights with considerations of economic efficiency in that the regulation must not require an operator to incur unnecessary expense.

---

[36] In legal terms, most states have adopted the view that the owner of a tract of land does not own the oil in place but acquires title by production from wells on his land. Other states, as Texas, confer title to the tract owner, but he loses title if the oil migrates from under the land. See Robert E. Hardwicke, "Rule of Capture," *Texas Law Review*, Vol. 13 (June, 1935), pp. 391–422.

The rule to provide adequate protection leads to the necessity of a well-spacing rule that will effectuate it. This is supplied in Rule 5.1 of the IOCC model statute as follows: "An order establishing spacing units shall specify the size and shape of the units, which shall be such as will, in the opinion of the Commission, result in the efficient and economical development of the Pool as a whole. The size of the spacing unit shall not be smaller than the maximum area that can be efficiently and economically drained by one well. . . ." Rule 5.2 is, "Except where circumstances reasonably require, spacing units shall be of approximately uniform size and shape for the entire Pool." Rule 5.2 would probably not be necessary if pools were unitized.

The subject becomes more complicated when, at the level of allocation among wells within a single reservoir, the IOCC model statute attempts to state an equitable principle. It says in Rule 4.2, ". . . the Commission shall, subject to the reasonable necessities for the prevention of Waste, allocate the allowable production among the several wells or producing properties in the Pool so that each Person entitled thereto will have a reasonable opportunity to produce or receive his Just and Equitable Share of the Production." Rule 1.1.12 states, " 'Just and Equitable Share of the Production' means, as to each Person, that part of the authorized production from the Pool that is substantially in the proportion that the amount of recoverable Oil or Gas or both in the Developed Area of his tract or tracts in the Pool bears to the recoverable Oil or Gas or both in the total Developed Areas in the Pool." And Rule 1.1.13 says, " 'Developed Area' means a spacing unit on which a well has been completed that is capable of producing Oil or Gas, or the acreage that is otherwise attributed to a well by the Commission for allowable purposes."

An effort to approximate this principle would imply a proration formula based on the recoverable reserves underlying each drilling unit in each pool. If the pool were relatively uniform throughout, this might be approximated by uniform well spacing. If characteristics differed in different parts of the pool, it might be approximated by a formula based on acre-feet of producing sand. Mere acre-feet takes into account variations in the thickness of the producing sand but does not take into account variations in porosity or permeability that may greatly influence the amount of oil in place and its recoverability. A formula based on surface acreage ignores variations in formation thickness as well. The IOCC statutory wording does not clearly cover special problems that would arise in connection with position of wells on the structure and different driving mechanisms, though it might be construed to do so. An effort to extend the equitable principle to these situations would presumably require unit operation of the pool. Probably the most difficult situation arises where a tract, because of its location on the structure, would, under normal depletion without regulation, recover several times the amount of reserves underlying the tract. How the owner of such a tract fares under regulation would depend upon the particular rule applied.

When these implications of the IOCC principle are brought out, it appears that no state attempts to follow any such concept of correlative rights consistently, though it is in a degree expressed in some judicial decisions.[37] Such approach to it as there is comes mainly through wide uniform well spacing and 100 per cent acreage or acre-feet formulas in proration. The degree of approach on these two points differs greatly from state to state. In Texas, Rule 37 exceptions have in the past commonly stood the IOCC principle on its head, conforming to the philosophy that the owner of any lease, no matter what size, should be entitled to drill a well and to recover enough oil to pay for the cost of drilling and return a profit on his investment. The "equitable" principle was that of licensing small-tract operators to drain oil from the property of their neighbors. Recent judicial decisions will presumably allow, if not force, the commission to abandon this practice.

Data collected for the current IOCC efficiency study indicate that in recent years state commissions have been moving toward 100 per cent acreage or acre-feet formulas, partly to comply with the law as announced by courts. In Texas, for example, the Railroad Commission set 100 per cent acreage formulas in 73 out of 100 new oil fields in 1962. This, however, is only a small dent in the older system where thousands of producing pools continue to operate under formulas having heavy well factors that in general express the small-tract philosophy. As noted above, even 100 per cent acreage formulas go only a limited distance toward fulfilling the IOCC definition of correlative rights. Many other factors might be relevant in particular cases.

[37] See p. 180, footnote 63.

Allocation among Pools

The IOCC model statute does not attempt to state a "fair share" principle with respect to the allocation of the statewide allowable production among pools. All it has to say on this point is in Rule 4.1, as follows: "Whenever the Commission limits the amount of Oil that may be produced in the State, the Commission shall allocate the allowable production among the Pools on a reasonable basis." Taking a practical view of the matter, this is probably as far as a statutory provision can sensibly go. It recognizes the necessity for regulatory agencies to take account of the special circumstances of different reservoirs. Some statutes instruct the regulatory authorities to allocate on a reasonable basis or without discrimination among pools. In practice this is likely to mean allocation mainly in line with market demand since pool output is on the whole governed by the proration formulas. In any case, the authorities have to establish some "reasonable" criteria by which to justify their treatment of pools.

The industry literature, if read uncritically, might lead to some misapprehensions about what happens. For example, William J. Murray, Jr., then a member of the Texas Railroad Commission, wrote: "*Proration* might be defined as the ratable and equitable allocation to each well in a given field of a total based upon either the maximum amount of petroleum which the reservoir characteristics of the field will permit it to produce efficiently, or that portion of the total market demand which may be properly assigned to the particular field, whichever quantity is smaller. This is known as MER and market demand proration. It is pointless to attempt one without the other, since it has been proved that conservation can only be accomplished by a combination of both types of proration."[38]

This *sounds* as though it meant that each reservoir has an MER rating and that, when it is necessary to prorate to market demand, each reservoir is reduced proportionally from this rated amount— the wells then sharing pro rata in the reduced total of the reservoir. This is presumably not what Mr. Murray meant, since it does not correspond to the practice in Texas of which he was one of the

---

[38] Wallace F. Lovejoy and I. James Pikl, Jr. (eds.), *Essays on Petroleum Conservation Regulation* (Dallas: Southern Methodist University, 1960), p. 68.

architects. In that case, he can mean only that pool output is governed by the proration formulas, except where special field rules are introduced.

That there is a misconception on the point is illustrated by a passage from a report by the Attorney General of the United States. In Texas, the report states,[39] "In general each well and field is assigned a 'maximum efficient rate' (MER) of production on the basis of engineering and other factors affecting the field. The wells capable only of a production rate below the legal definition for marginal wells are allowed whatever production they can make. The monthly allowable is then set at that portion of the month's production at MER required to meet the market demand, less the amount produced by the marginal wells." This is not what happens. The report confuses MER with the poolwide total under the depth-factor schedule.

Erich W. Zimmermann's statement of a principle of allocation adds to the confusion: "The overruling principle of just, nondiscriminatory allocation applies both to allocation of pool allowables among properties and to the allocation of the state allowable among pools. In order to accomplish the former, one must almost necessarily do the latter. But allocation of production quotas for purposes of conservation must not only serve the protection of property rights but contribute toward waste elimination. The elimination of waste requires that no pool be 'produced' at a rate in excess of its MER. The allocation of the state allowables among pools, therefore, cannot be made on the basis of total recoverable reserves but to a considerable extent must be made on the basis of MER's at which these reserves may currently be produced if underground waste is to be avoided. . . . an equitable allocation policy must take into account that waste prevention requires that different pools must be 'produced' at different rates of production and that therefore it takes a longer time to produce the quota of some pools than of others."[40] Whether Professor Zimmermann means that the equitable principle among pools is to cut each pool back proportionally to its MER, we

---

[39] U.S. Department of Justice, *Second Report of the Attorney General Pursuant to Section 2 of the Joint Resolution of July 28, 1955, Consenting to an Interstate Compact to Conserve Oil and Gas* (Washington, 1957), p. 85. See also p. 116 of report.

[40] Erich W. Zimmermann, *Conservation in the Production of Petroleum* (New Haven: Yale University Press, 1957), pp. 330–31.

were unable to decipher. If so, what happens does not conform to the rule. As we noted earlier, Texas has vacillated back and forth between a yardstick and MER's as a basis for allocation, and Zimmermann does not make this clear.

There is no obvious and simple principle of equity by which to determine the allocation of statewide production among reservoirs. In this situation, administrative convenience and feasibility take over. The regulatory agencies solve the problem of equitable allocation among reservoirs by evading it. They set up formulas to regulate production per well, regardless of the reservoir in which wells are located. Production per reservoir is simply the composite result of production under the per-well formula, as modified by field orders to take account of special circumstances. In some instances departures can be found, such as basing production on MER's. Even in unitized fields the yardsticks and proration formulas usually dictate the pool allowable, again with some modifications in some instances. The results of all this may be considered "reasonable" insofar as the top-allowable schedules can be considered reasonable. The principal feature of the top-allowable system is the depth-acreage formulas. The different allowables for different depths do not of themselves raise any question of correlative rights. Of two reservoirs having similar characteristics except depth, the deeper one will be produced more rapidly. The discrimination is simply one of timing of returns, which, of course, may be substantial. The justification of this discrimination is one of expediency—the desire to stimulate deeper drilling. It is a matter of judgment how great the depth differential should be. Of two reservoirs at the same depth but differing greatly in drilling costs, there is discrimination in favor of the less costly in that its payout period will be much shorter.

Unallocated Pools

Another phase of proration where one might look for equitable considerations is the classification of wells and reservoirs as "allocated" and "unallocated." The usual stated reason for placing so-called "marginal" wells and reservoirs in the unallocated category is to "prevent premature abandonment," which could serve as a principle both of fairness and of efficiency in terms of ultimate recovery. In the formal language of its July, 1963 proration order, "Findings," the Oklahoma Commission says that ". . . the Commission has classified each of the common sources of supply in the State of Oklahoma into divisions known as allocated and unallocated pools; that such classifications are necessary and proper in order that waste be prevented and that there can be no discrimination between the various sources of supply . . ." As against this formal language, two better reasons could probably be assigned in most market-demand proration states: (1) that attempting to apply the quota system to tens of thousands of small wells would be administratively burdensome but not impossible; and (2) that any attempt to cut low-yield wells back to lower rates would arouse intense political opposition. Every state has its own practices of inclusion in the unallocated list, and in Texas and Kansas much of it is fixed by statutory fiat. There is no obvious equitable principle of fairness at work, except that of providing a little more current income to a great many people. At the bottom level, in the region of abandonment, the rules undoubtedly delay abandonment, but it would not necessarily be premature in economic terms.

At the end of this brief survey, it appears that there is no clearly defined concept of fair shares or nondiscrimination, either for producers within pools or as between pools. Each state regulates access of private owners to gain from their oil-bearing properties by its own statutory and regulatory rules, mainly through well-spacing rules and depth-acreage proration formulas. No state appears to conform at all closely to the definition in the IOCC model statute, and the bias of the system in most states is in an opposite direction, insofar as proration rules are weighted by a per-well factor. A 100 per cent acreage or acre-feet formula operates in the direction of the IOCC definition, but by no means approximates it.

Under present methods of reservoir development, it is obvious how difficult it would be to assign quotas to drilling units closely corresponding to the relative amounts of recoverable oil in them at minimum cost. Some approximation is possible through wide and equal well spacing. The problem in IOCC terms could perhaps be solved with greater accuracy under unit operation, if placed in the hands of engineers who made their findings on the basis of reservoir geography and geology. The present system, according to its specific rules, can be strongly biased toward one group of interests or another, but solutions here have typically been found through the arts of compromise or "force majeure," and are not easily reducible to principles of fairness. In the degree that current developments run in the direction of wide and uniform spacing and 100 per cent acreage or acre-feet formulas, the change can probably be attributed mainly to the pressure of economic cost factors, but partly also to some deliberate reconsideration of equitable adjustment of property rights.

5. COMPETITION AMONG STATES FOR MARKETS

In the early consideration of an interstate compact between the oil-producing states, it seems to have been anticipated that the compact authority would determine a nationwide production limit and allocate the total among the various member states, to be further allocated by them to their producers. This feature would have involved some participation by the federal government because of its control of imports. As we have seen, however, no such provision was included in the compact that created the IOCC, with the result that each state remained solely responsible for regulating the amount of its own production.

Each state is naturally interested that its own production should be as large as possible, consistent with market stability and "waste prevention." Under the dynamics of the developing industry, however, from stage to stage some states forged ahead while others fell behind, and those at the lower end usually have felt dissatisfaction with the state of affairs. These feelings have led to intermittent proposals that the IOCC should set up procedures for allocating among the states or at least to study the question of "fair shares" among states.

The most recent outcrop of such sentiments led in February, 1960, to the appointment of a special IOCC committee on "Study of Equality of Opportunity of Oil Production by the Sovereign States." After meetings in which serious differences of view developed among representatives of various states, it was first changed into a subcommittee of the regulatory practices committee; thereafter, its duties were taken over by the executive committee, composed of governors of several states. After unreconcilable differences of view developed, discussion of the subject was abandoned.

The Governors' Committee of the IOCC in 1964 took a hard look at state market shares but included very little of its analysis in the final report. This was presumably because of the substantial disagreement among states in the area. The final report concluded that, "The interests of the consumer have been and can be served best by the existing interplay of free competition, not by any allocation of total national demand."[41] This study, however, did present some interesting comparative production and reserves data, some of which are included in Table 12. This table shows the percentage of liquid-hydrocarbons production and reserves that each of the seven largest producing states had in 1952 and 1963.

TABLE 12. PERCENTAGE OF U.S. LIQUID-HYDROCARBONS
PRODUCTION AND PROVED RESERVES, BY STATES,
1952 AND 1963

State	Per cent of production		Per cent of reserves	
	1952	1963	1952	1963
Texas	45.4	37.8	56.5	47.7
Louisiana	10.7	18.3	9.2	17.5
California	15.4	10.5	12.7	10.2
Oklahoma	8.3	7.6	5.6	5.4
Wyoming	2.8	4.5	3.2	3.6
New Mexico	2.6	4.4	2.3	4.1
Kansas	4.7	3.8	3.0	2.7
Total: 7 States	89.9	86.9	92.5	91.2

SOURCE: IOCC, Report of the Governors' Special Study Committee. *A Study of Conservation of Oil and Gas in the United States* (Oklahoma City, 1964), pp. 107–40.

The production facts underlying the controversy among the states may be seen in Table 13 for PAD Districts I to IV. From the table it is seen that from 1956 to 1962, there was an increase of 298,400 barrels per day which was, however, the net of increase of 909,900 barrels per day in eighteen states and decreases of 611,500 barrels per day in nine other states. Of the improving states, Louisiana was the greatest gainer with a rise of 505,500 barrels over the six-year period. Wyoming was the second largest gainer, and important gains were also registered by Utah, New Mexico, Mississippi, and North Dakota—to take only those gaining more than 30,000 barrels per day. In the aggregate, the improving states increased their produc-

---

[41] IOCC Governors' Special Study Committee, *op. cit.,* p. xxi.

TABLE 13. CRUDE-OIL PRODUCTION BY STATES,
1962 COMPARED WITH 1956,
DISTRICTS I–IV

(thousands of barrels daily)

State	1956	1962	Net change: 1962 over 1956	Cumulative net change
States showing production increases in 1962 over 1956 levels				
Louisiana[a]	818.1	1,323.6	505.5	505.5
Wyoming	286.4	397.7	111.3	616.8
Utah	6.7	84.8	78.1	694.9
New Mexico[a]	240.1	297.8	57.7	752.6
Mississippi	111.5	149.2	37.7	790.3
North Dakota[a]	36.8	68.9	32.1	822.4
Montana	59.5	86.7	27.2	849.6
Nebraska	44.3	68.1	23.8	873.4
Michigan[a]	29.3	46.9	17.6	891.0
Alabama[a]	8.4	20.5	12.1	903.1
West Virginia	6.0	9.2	3.2	906.3
Kentucky	48.2	49.6	1.4	907.7
Ohio	13.1	13.9	0.8	908.5
Indiana	31.5	32.1	9.6	909.1
4 other states[b]	—	0.8	0.8	909.9
States showing production decreases in 1962 over 1956 levels				
Texas[a]	3,026.8	2,565.8	(461.0)	(461.0)
Oklahoma[a]	589.8	544.4	(45.8)	(506.8)
Colorado	159.9	116.3	(43.6)	(550.4)
Kansas[a]	339.4	307.1	(32.3)	(582.7)
Illinois	225.0	211.8	(13.2)	(595.9)
Pennsylvania	22.5	14.3	(8.2)	(604.1)
Arkansas[a]	80.2	75.6	(4.6)	(608.7)
New York	7.5	4.9	(2.6)	(611.3)
Florida[a]	1.3	1.1	(0.2)	(611.5)
Total states in Districts I–IV				
	6,192.3	6,490.7	298.4	298.4

[a] Market-demand state.
[b] Combined data for Missouri, South Dakota, Tennessee, and Virginia.
SOURCE: U.S. Department of Justice, *Report of the Attorney General Pursuant to Section 2 of the Joint Resolution of August 7, 1959, Consenting to an Interstate Compact to Conserve Oil and Gas* (Washington, 1963), p. 22 (mimeo). 1956 production data as reported in 1957 *Minerals Yearbook;* 1962 production data as reported in December, 1962, *Monthly Petroleum Statement* issued by the U.S. Bureau of Mines.

tion by 52 per cent and they raised their proportion of total production in Districts I–IV from 25 per cent to 41 per cent.

Of the losers, Texas was hardest hit with a loss of 461,000 barrels per day, or 15 per cent, but Oklahoma, Colorado, and Kansas were also hard hit. As we noted earlier, the percentage figure for Texas does not show how severe the strain was on its regulatory system, since much of its production is exempt from allocation, and the cutbacks have to be charged against the nonexempt portion. Something similar is true of Oklahoma and Kansas.

For a trend view, 1952 provides a better base than 1956, since the years 1952–62 ride over the Suez peak of 1956–57 as well as over the business recessions of 1953–54 and 1958–59. Using the 1952 base, in the ten-year period, Texas production declined 7.2 per cent; Kansas declined 3.4 per cent; Oklahoma increased 3 per cent. On the other side of the picture, Louisiana increased 100 per cent; New Mexico increased 85 per cent; all other states in Districts I–IV, taken together, increased 78 per cent (though this includes some small decreases). The basis for these calculations is shown in Table 14.

Such figures demonstrate how heavily the difficulties overtaking producers and regulators have been concentrated in the states of Texas, Oklahoma, and Kansas, where crude oil has historically occupied so important a place in the economy of the state. As we shall see in a moment, the producers and regulators in these states are caught in a squeeze that is partly of their own making.

The reasons for the shifting patterns of production and reserves can perhaps be generalized into the overly simple statement that oil is developed, produced, and purchased where it is most profitable. Profitableness in turn depends in part on some natural phenomena such as the suspected or actual presence of prolific reservoirs and geographic location relative to markets. It also depends in part upon relative economic factors including regulations that affect both costs and revenues.

When the special IOCC committee was set up in 1960, it seems possible that the problem was thought of as mainly an issue between states that practiced market-demand proration and those that did not, since the committee was originally constituted with equal membership from these two groups of states. Certainly, as discussion proceeded, one of the controversial issues that arose was the feeling of the "non-market-demand" states that an effort was being made to coerce them into restrictions that they saw no reason to impose.

However, as Table 14 demonstrates, the gains and losses were not at all distributed according to the market-demand basis of regulation. Both the greatest gainer—Louisiana—and the greatest loser—Texas—were market-demand states, and both groups are well represented among gainers and losers.

When an issue of this sort is drawn, it is natural that states on opposite sides of the gain–loss divide should attempt to advance "equitable" arguments in support of their own interests. Except for Louisiana, we have not seen briefs prepared by the affected states. Louisiana, however, provides a good case in point. In a statement made by Governor Davis in comment upon the figures in the Attorney General's *Report*,[42] he points out that Louisiana's proportion of the U.S. production prior to 1960 had run substantially below its proportion of U.S. reserves and makes the equalization of these proportions an "equitable" argument for Louisiana's position.[43]

Similar, and even stronger, arguments are available to Texas. In a statement to the Texas Railroad Commission in July, 1961, speaking of the plight of independent producers due to declining production, the president of the Texas Independent Producers and Royalty Owners Association (TIPRO) noted that "Texas now has 53 per cent of the proved reserves in Districts I–IV and only 41 per cent of its production. It holds approximately two-thirds of the nation's excess producing capacity and could double its present production rate." Texas can bring both reserves and capacity to support its "equitable" argument. One can question the arguments of both Louisiana and Texas. Should a state necessarily have a share of the national production or market that is equal to that state's share of national producing capacity or reserves? This hardly seems rational. Kuwait in 1964, for example, had reserves estimated at about 63 billion barrels and daily production of about 2.1 million barrels.

[42] James H. Davis, Governor of Louisiana, "Louisiana's Position with Respect to the Report of the Attorney General, Dated May 15, 1963, Pursuant to Section 2 of the Joint Resolution of August 7, 1959 Consenting to an Interstate Compact to Conserve Oil and Gas," presented to the Chairman of the Interstate Oil Compact Commission, May 23, 1963.

[43] From 1949 through 1954 Louisiana's share of national production of liquid hydrocarbons (crude oil and natural-gas liquids) exceeded slightly its share of national reserves of liquid hydrocarbons. From 1955 through 1961 Louisiana's position reversed; its share of reserves exceeded slightly its share of production. In 1962 and 1963 the position once again reversed, with the share of production exceeding the share of reserves somewhat. IOCC Governors' Special Study Committee, *op. cit.*, p. 154.

TABLE 14. ESTIMATED DEMAND FOR DOMESTIC CRUDE OIL ORIGINATING IN DISTRICTS I-IV, REFINERY RUNS OF FOREIGN CRUDE OIL IN DISTRICTS I-IV, AND REFINERY INPUT OF NATURAL GAS LIQUIDS IN DISTRICTS I-IV, 1946-62

	Demand for domestic crude oil									Refinery runs foreign crude oil	Refinery input natural gas liquids	Total demand (cols. 9+10+11)
	Market-demand states (state of origin)								Total domestic crude			
Year	Texas	Louisiana	Oklahoma	Kansas	New Mexico	5 other states[a]	Sub-total	Other states				
	(1)	(2)	(3)	(4)	(5)	(6)	(7)	(8)	(9)	(10)	(11)	(12)
	Quantities in thousands of barrels daily											
1946	2,074.6	397.4	383.2	265.1	100.0	124.5	3,344.8	540.1	3,884.9	231.1	125.8	4,241.8
1947	2,220.7	439.3	395.5	291.0	112.0	127.8	3,586.3	593.5	4,179.8	266.5	135.1	4,581.4
1948	2,454.0	490.2	419.8	299.5	129.4	133.7	3,926.6	613.9	4,540.5	338.8	149.1	5,028.4
1949	2,062.8	519.2	412.5	281.9	128.9	130.3	3,535.6	616.5	4,152.1	423.6	165.9	4,741.6
1950	2,254.4	574.0	450.7	290.6	132.7	135.6	3,838.0	1,153.4	4,991.4	482.2	186.4	5,660.0
1951	2,757.9	634.6	510.1	312.6	143.0	123.1	4,481.3	672.4	5,153.7	488.1	205.1	5,846.9
1952	2,763.9	661.5	524.5	318.6	160.5	123.9	4,552.9	691.0	5,243.9	532.2	213.5	5,989.6
1953	2,805.2	701.0	560.0	316.1	190.6	134.6	4,707.5	757.7	5,465.2	567.1	229.3	6,261.6
1954	2,701.7	682.1	513.9	324.7	206.8	135.0	4,564.2	840.4	5,404.6	603.6	243.0	6,251.2
1955	2,862.2	738.7	562.6	333.7	224.4	146.1	4,867.7	944.6	5,812.3	681.2	266.8	6,760.3
1956	3,049.1	812.0	586.0	334.4	242.4	151.4	5,175.3	1,016.9	6,192.2	755.3	295.9	7,243.4
1957	2,916.1	902.3	600.2	339.6	256.5	166.3	5,181.0	1,047.5	6,228.5	739.4	340.5	7,308.4
1958	2,599.0	868.2	553.2	328.9	271.6	162.9	4,783.8	1,106.0	5,889.8	757.5	310.5	6,957.8
1959	2,680.2	990.5	540.1	328.4	288.3	164.4	4,991.9	1,212.5	6,204.4	727.1	354.6	7,286.1
1960	2,562.8	1,089.1	530.9	310.9	294.1	207.0	4,994.8	1,240.9	6,235.7	738.3	384.1	7,358.1
1961	2,556.7	1,163.7	526.8	310.2	302.0	216.9	5,076.3	1,241.9	6,318.2	741.3	393.9	7,453.4
1962	2,564.6	1,315.7	540.5	307.6	299.3	214.5	5,242.2	1,236.3	6,478.5	769.0	435.2	7,682.7

Per cent of total demand set forth in Column 12 above

Year												
1946	48.91	9.37	9.03	6.25	2.36	2.93	78.85	12.73	91.58	5.45	2.97	100.0
1947	48.47	9.59	8.63	6.35	2.45	2.79	78.28	12.95	91.23	5.82	2.95	100.0
1948	48.80	9.75	8.35	5.96	2.57	2.66	78.09	12.21	90.30	6.74	2.96	100.0
1949	43.50	10.95	8.70	5.95	2.72	2.75	74.57	13.00	87.57	8.93	3.50	100.0
1950	39.83	10.14	7.96	5.13	2.35	2.40	67.81	20.38	88.19	8.52	3.29	100.0
1951	47.17	10.85	8.72	5.35	2.45	2.10	76.64	11.50	88.14	8.35	3.51	100.0
1952	46.14	11.04	8.76	5.32	2.68	2.07	76.01	11.54	87.55	8.89	3.56	100.0
1953	44.80	11.20	8.94	5.05	3.04	2.15	75.18	12.10	87.28	9.06	3.66	100.0
1954	43.22	10.91	8.22	5.19	3.31	2.16	73.01	13.44	86.45	9.66	3.89	100.0
1955	42.34	10.93	8.32	4.94	3.32	2.16	72.01	13.97	85.98	10.07	3.95	100.0
1956	42.09	11.21	8.09	4.62	3.35	2.09	71.45	14.04	85.49	10.43	4.08	100.0
1957	39.90	12.35	8.21	4.65	3.51	2.27	70.89	14.33	85.22	10.12	4.66	100.0
1958	37.35	12.48	7.95	4.73	3.90	2.34	68.75	15.90	84.65	10.89	4.46	100.0
1959	36.78	13.59	7.41	4.51	3.96	2.26	68.51	16.64	85.15	9.98	4.87	100.0
1960	34.83	14.80	7.22	4.23	4.00	2.81	67.89	16.86	84.75	10.03	5.22	100.0
1961	34.30	15.61	7.07	4.16	4.05	2.91	68.10	16.66	84.76	9.95	5.29	100.0
1962	33.38	17.13	7.04	4.00	3.90	2.79	68.24	16.09	84.33	10.01	5.66	100.0

a The five other market-demand states are Alabama, Arkansas, Florida, Michigan, and North Dakota.

SOURCE: U.S. Department of Justice, *Report of the Attorney General Pursuant to Section 2 of the Joint Resolution of August 7, 1959, Consenting to an Interstate Compact to Conserve Oil and Gas* (Washington, 1963), pp. 30–31 (mimeo). 1956 production data as reported in 1957 *Minerals Year-book*; 1962 production data as reported in December, 1962, *Monthly Petroleum Statement* issued by the U.S. Bureau of Mines.

The United States had reserves of 34.5 billion and daily production of 7.6 million barrels. Economic factors seem much more relevant in determining market shares.

While the mutual interest of the states in market stability is obviously a major factor in making the system work, no single state has any incentive to curtail its production below that for which it has a ready market. The purchasers, at least indirectly, decide this, spreading their purchases among states as they see fit. When a state like Louisiana is on the upgrade and a state like Texas is on the downgrade, it is not *because* the Louisiana commissioner increases allowable production and the Texas Railroad Commission does the opposite. On the contrary, both are controlled by what is saleable at the going price. Fortuitous circumstances within the dynamics of the industry and the policies of large purchasers determine this. Taking a short-run view, state regulatory commissions do not in a direct sense effectually determine the amount of production in a state. They spread out the salable amount among the various producers. Indirectly, however, and taking a longer-run view, the several state regulatory commissions and the state laws under which they operate can and do affect the relative attractiveness of operating in one state rather than another. As we have said before, a major integrated company that both purchases and produces oil must be governed by profitability at least in the long run. To do otherwise makes no sense as a matter of business practice.

There appears to be a partial misconception about the role of Texas, which is sometimes referred to as the "balance wheel" of the market stabilization process, suggesting that it is an exception to the preceding statements. A trace of the misconception shows up in De Chazeau and Kahn in the following passage:

The Texas Railroad Commission which controls more than 40 per cent of the United States crude production and approximately half of estimated national proved reserves, has also found it desirable to make allowances in its estimates for anticipated production in states that do not control production or use other bases of control. By thus allowing for estimated supplies beyond its jurisdiction, Texas in effect brings total available supply, including imports, within the principle of prorationing to market demand.[44]

---

[44] M. G. de Chazeau and A. E. Kahn, *Integration and Competition in the Petroleum Industry* (New Haven: Yale University Press, 1959), p. 123.

This is not quite what happens, or put more precisely, the reasoning attributed to the Texas commission seems to be found in all the major market-demand states. Texas, being the largest, clearly plays an important role in the stabilizing process. We shall pursue this area of discussion in Chapter 8. Texas, like all other states, authorizes the production of all the oil the commission thinks its producers can sell, consistent with stable prices and with standards of sharing among the various producers. What Texas and, to some extent, all the market-demand states must live with is a long history of regulation that places substantial shackles on regulatory policy. *If* Texas modernized its system and made the industry more efficient by its regulations, it is quite conceivable that Texas could increase its allowables even in the face of stable national demand. Purchasers will buy and operators will drill where the greatest profits can be made. One effect of overcapacity is that it creates pressures in every state to reduce costs. A competitive race among states to see which could establish regulations that generated the lowest costs for its industry would indeed be a pathbreaking event.

Professor Zimmermann, though he uses the term "balance wheel" somewhat ambiguously, on the whole describes the matter correctly.[45] The idea that Texas is the sole balance wheel is no doubt a holdover from the early days of proration when severe cutbacks by Texas played a major role in raising prices and achieving relative stability in markets. The system of stabilizing influences having become fully operative among states generally, the principles of behavior are the same for Texas as for everyone else. It should also be noted that there is another major influence present but inactive. This is the federally controlled import quota system which governs a larger share of total supply than any single state except Texas. Whether it will continue to be dormant remains to be seen.

It has been suggested by some observers that the Texas market-restriction mechanism of "shutdown days" was partly responsible for the loss of markets by this state. Prior to January, 1963, the Texas commission issued a monthly shutdown order which contained language to the effect that ". . . each and every oil well in each and every field . . . except as hereinafter provided shall be shut in during said period of time a total of twenty-three (23) days,

---

[45] Zimmermann, *op. cit.*, pp. 213–14.

giving to all of said wells in said fields a total of eight (8) producing days during said thirty-one (31) day period."[46] The equating of production to consumption needs was thus stated in terms of number of days production at yardstick rates. If the commission estimated that purchasers would be willing to take 8½-days production, it had the choice of setting rates at either 8 days or 9 days; it would not use fractional days. It is claimed that the choice invariably was for the lower number of days rather than the higher because of the commission's fear of creating a surplus that might have disruptive market effects. If, as some people claim, the commission consistently behaved in this fashion, the result could conceivably have been a shrinking of the market for Texas crude, since purchasers would never quite fill their needs. Such behavior might be explained by arguing that the commission had a balance-wheel philosophy and saw its role as maintaining stability in the national crude oil market.

In January, 1963, the commission changed the restriction mechanism by discarding the clumsy shutdown-day system and adopting a percentage-of-yardstick system similar to that used in virtually all other market-demand states. This certainly provides a degree of fineness that the old system never had. If this entire line of reasoning is correct, we are at a loss to understand why Texas did not change its method years ago.

It may be added that if Texas wished to destroy the whole proration system by a prolonged deliberate increase of production, it could do so with ease and dispatch. The same could be said for Louisiana or Oklahoma. There is no likelihood that Texas would attempt to recoup its position by so drastic a method. All states have a common interest in market price stability and would be damaged by the unleashing of production that would seriously undermine the price structure. Over a long period of years no state or group of states has followed practices that would lead to this result. The only way this could be done would be to authorize and encourage production well beyond the existing market demand at current prices in one or more important producing states, thereby creating price competition among producers. In guarding its own interests, no state is likely to resort to such a practice. If it attempted to do so, it would merely upset its own applecart. Price competition,

---

[46] Order No. 20–50,010, November 15, 1962. The number of shutdown and producing days varied from month to month.

once it had led to a general price decline, would probably not result in much redistribution of demand among the states, but would simply leave them all producing about the same amount as before, with a lower return in the short run to the producers and to the state tax coffers. If such moves carried the whole regulatory structure away, no one could predict the outcome.

Within the limits of necessary self-restraint, every state is a competitor for a share of the crude-oil market, and its behavior is, therefore, dictated by its own conception of its own interest. At any given time, the balance of advantage is running with some states and against others. It is difficult to imagine that any state in the former group would voluntarily forego its advantages, except under the ultimate threat that the system would collapse if it did not. Since no state will wish to use this ultimate sanction—undertaking the role of Samson—no state is under compulsion to act other than in its own interest. This is why voluntary agreements to share the total market among the states have made no progress.

When looked at in the aggregate for the whole country, crude-oil-production statistics give the impression of a relatively static industry. But when broken down regionally and by states, the picture is one of highly dynamic development in some areas and of the opposite in others. Over the next few years, the states in the relatively declining group will have to examine their situations to determine how far their decline is due to geological and economic causes beyond their control, and how far it may be due to their own regulatory policies which they have the power to change.

However astute its analysis of the difficulties, a state regulatory agency will never find it easy to make the necessary adjustments because of the effect on vested interests. This is especially true of Texas, and may be illustrated from the attitude of TIPRO, the most active agent of the independent producers. In the statement referred to earlier (p. 223), the president of TIPRO said: "Interstate oil purchasing companies are failing to cooperate with producing states in sharing equitably the nation's reserve producing capacity. Clearly this is the primary cause of today's market sharing problem." The motivation of major integrated companies is interpreted as a desire to favor their own production, which leads them out of Texas where so much of the production is in the hands of independents.

In the short run, it is difficult to see why the majors should increase their purchases from Texas independents when oil is available to them on better terms elsewhere. In the longer run, the problem merges into the more general problem of introducing economic efficiency into the structure of the industry. The kinds of measures necessary to effect this result—wide well spacing, unit operation, elimination of high-cost "marginal" wells, higher allowables to low-cost sources—have not heretofore had the support of many independent producing interests. The question to which TIPRO appears not to have faced up is how the relative position of Texas can be expected to improve without radical changes in the regulatory practices that are in a substantial degree responsible for the difficulties.

Their statement of the problem has at least the merit of introducing an element of realism about attitudes within the industry. While the conflicting interests of states are "real," a deeper reality is the conflict of interest within the industry itself. One question that faces state regulatory agencies is where they will throw their weight in this internecine industry conflict and how much power they have to affect the outcome by policy expedients of the stick or carrot variety.

6. STATISTICAL SERVICES OF THE
FEDERAL GOVERNMENT

The statistical services provided by the U.S. Bureau of Mines play an important part in state petroleum conservation efforts. While the bureau publishes tremendous amounts of statistical data on all mineral industries, its work with petroleum information is probably the most comprehensive. These data are used extensively not only by state regulatory authorities but also by all the segments of the industry, by federal policy making groups, and by anyone studying this industry. Generally speaking, the bureau's statistical reports contain data that are of a primary nature. In most instances very little interpretive text accompanies the data, and virtually no analysis of trends or variables is found. Thus, these services can properly be called informational rather than analytical and are provided for anyone's use. From time to time criticism is heard of these services, usually along the lines that the data are misused or are used to manipulate prices, production, or some segment of the industry. Such criticism is hardly justified since the bureau cannot be held responsible for how the information it provides is used. Over the years these services have been expanded and improved to the point where no other country has as fine a system and, within this country, probably no other industry has better or more current information at its disposal. Of course, it is not ideal.

Of particular concern to state conservation agencies are two sets of data: the monthly petroleum forecasts and the weekly and monthly reports on above-ground stocks or inventories of crude oil, natural gas liquids, and refined products. Both of these series, and particularly the latter, are used by market-demand states to assist in setting state allowables. Other data reported by the bureau are significant but are used to obtain a general view of the conditions of the industry. Included here are such things as refinery runs, refinery yields, refinery outputs, refinery capacity, shell capacity of storage facilities for products, crude oil and NGL production, domestic movements of crude oil, NGL and refined products, imports and exports of crude oil and refined products, and several others. This information is often reported by state, refinery districts, PAD districts, or in other geographical detail.

Monthly Petroleum Forecasts

Perhaps the most controversial statistical service of the bureau is its "Forecast of Market Demand (Consumption) of Crude Oil,"[47] which usually appears late in one month for the following month. This is a short-term forecast of estimated consumption of crude oil by states. Any state proration agency requesting it can receive an advance release of this forecast in time for its monthly (or bimonthly) market-demand hearings. Virtually all the market-demand states receive this information. There are several sources that describe these forecasts, their use, and their accuracy in considerable detail.[48]

---

47 In U.S. Bureau of Mines, *Monthly Petroleum Forecast.*

48 National Resources Committee, *Energy Resources and National Policy,* "Bureau of Mines Forecasts of Demand for Motor Fuel and Crude Oil," by Alfred G. White (1939), pp. 401–14; hearings before the Temporary National Economic Committee, *Investigation of Concentration of Economic Power,* Part 17, *Petroleum Industry,* 76th Cong., 2d Sess. (1939), testimony of Alfred G. White, pp. 9583–603; hearings before the Special Committee to Study Problems of American Small Business, Part 27, *Oil Supply and Distribution Problems,* U.S. Senate, 80th Cong., 2d Sess. (1948), testimony of Alfred G. White, pp. 3138–76; U.S. Attorney General, *Second Report of Attorney General on the Interstate Compact to Conserve Oil and Gas, op. cit.,* pp. 33–49.

The history of these forecasts goes back as far as 1930 when the Federal Oil Conservation Board, chaired by the Secretary of the Interior, appointed a committee to make forecasts of gasoline and crude-oil demand. Several such forecasts were made in 1930, 1931, and 1932, with the cooperation of the Bureau of Mines. With the advent of the National Recovery Administration (NRA) in 1933, and the establishment of the Petroleum Administration Board (PAB), the Bureau of Mines began furnishing statistics to the PAB for use in determining oil allowables under the NRA petroleum code. This continued until the demise of the NRA in May, 1935. Up to this time the bureau made forecasts of demand (consumption) and of changes in levels of stocks, so that, in effect, the forecast was one of permitted production by states.

By 1935, several major oil-producing states had passed conservation statutes and were in a position to take over some of the allocation functions performed by the PAB under the petroleum code, although there was an obvious change in the legal status of the allocation system. The governors of several states requested that the bureau continue making demand forecasts to assist the states in conservation regulation. This was started in June, 1935. The Interstate Oil Compact Commission, formed in September of that year, seconded this request but asked that the bureau cease making forecasts of changes in stock levels, and this was done in November, 1936. Subsequently, the bureau has made only forecasts of demand (consumption) and has left to the discretion of the various state agencies the estimation of stock changes and thus the necessary production levels to equate supply and demand. Since November, 1936, this monthly forecast by states has been made, although from January, 1942, through October, 1945, only the national totals were published. At about the same time, November, 1936, the bureau started publishing past stock-level information in sufficient detail so that state agencies could make their own estimates of future stock levels and seasonal swings in stocks.

While these are called demand forecasts, they are more properly construed as consumption forecasts, since there is no effort to forecast what consumption might be under various sets of assumptions. The forecast is of what will be consumed, given existing conditions and levels of prices; facilities; capacities to produce, transport, and refine; and ability of consumers to use crude oil. The monthly

"Forecast of Market Demand (Consumption) for Crude Oil" contains the following statement (identical each month):

The forecast contains no recommendations as to desirable rates of crude production or of changes in crude stocks. It is an estimated or probable consumption based on actual industry conditions and policies in effect; adjustment for seasonal changes in refined stocks, allowances for the current rate of imports of foreign crude oil and products; and the production of natural gas liquids.[49]

The bureau has consistently refused to estimate what consumption might be under other than existing conditions. The use of this sort of definition of demand has raised many questions among critics of state and national fuels policy agencies. Under this concept it is quite conceivable that a situation might arise in which consumers of petroleum products cannot satisfy their demands at current prices because of a lack of transportation or refining capacity. This, in fact, was the case in 1947 and 1948 in some areas. This might even be true if there was excess producing capacity in the oil fields. The bureau does, in its forecast, discount any government policies that might affect demand, e.g., gasoline rationing, and will take into account, to the extent they are foreseeable, such things as interruptions in distribution channels due to strikes, emergencies, and abnormalities in weather. The bureau obviously does not attempt to make weather forecasts, but weather abnormalities of past months may influence significantly the demand for crude oil next month.

The mechanics of this forecasting are rather complicated in that the demand for crude oil is derived demand, and the initial consumption of crude oil by refineries may not take place in the state in which crude oil is produced or in the state in which the finished products are ultimately consumed. The bureau first starts at the national level and determines the demands for finished products. Gasoline, which now constitutes about 45 per cent of a refined barrel of crude, receives the most attention. The demand for gasoline, while having significant seasonal swings, is closely related to the number of motor vehicles being driven. Data on motor-vehicle registrations are easily obtained, and correlation of the number of vehicles registered with total motor fuel consumption yields fairly reliable figures on gasoline consumption per vehicle. Weather abnormalities can

---

[49] *Op. cit.*

cause substantial swings in month-to-month consumption, but the deseasonalized trend is quite clear. Much the same type of analysis is possible for heating oils—the second most important refined product. Data are available on the number of oil-burning heating units in operation and on the number of new installations of this type of equipment. Trends are relatively easy to discern although, as in the case of gasoline, seasonal variations may be quite large.

With these types of information on gasoline, heating oils, and other products, it is possible to get a good estimate of what consumption will be in the near future under normal weather and other conditions. It is also possible to convert product demand into demand for crude by projecting refinery yields from recent past months and from past seasonal variations in yield. Here, however, stocks of both products and crude become highly significant. If current stocks are high because of abnormal weather, next month's consumption of products may be quite normal but next month's consumption of crude oil may be below normal. This would merely reflect the fact that refineries are supplying their markets partly out of stocks and are not buying as much crude for current processing. Similar examples could be drawn for cases in which crude oil stocks are above or below normal.

The bureau has historical data that show the location and type of refining capacity, product yields of these refineries, and crude runs to stills. Given this information, plus an estimate of the national demand for crude oil derived from product demand, plus information on crude-oil and refined-product stock levels, it is possible to estimate crude oil demand by refinery district.[50] Past data on which crude-producing areas supplied how much oil to which refinery district, how much imported crude was processed in each, plus information on seasonal swings in runs to stills in various refinery districts enables the bureau to estimate the demands for crude oil by states. No attempt is made by the bureau to obtain any greater geographic breakdown, i.e., by railroad commission district in Texas or between southeast and northwest New Mexico.

While there are substantial variations between what is forecast and what is actually consumed for a given state in a given month,

---

[50] The bureau divides the United States into thirteen refinery districts. See *Minerals Yearbook, 1963, Vol. II, Fuels* (Washington: U.S. Government Printing Office, 1964), pp. 396–97.

the forecasts are fairly accurate. They are certainly accurate enough to provide guidelines for any state commission wanting to use them.[51]

A major criticism of the forecasts has been that they establish production quotas for producing states and at times keep demand estimates, and therefore production, below actual demand so that prices are made to rise or are kept at high levels.[52] It has even been claimed that certain interests control the forecasts in such a way as to keep prices high. These claims seem rather farfetched. While it is true that the bureau's forecasts are one piece of information considered by a state proration agency, it is not true that the states accept these estimates as the final, unquestioned word. The states, in fact, have perhaps much better estimates of demand, since they have purchaser nominations in their hands. Also, the demand for oil in a state may be supplied from either current production or stocks. When the evidence is reviewed, the conclusion must be that the market-demand states do not "slavishly follow" bureau forecasts.[53] It is our feeling that too much significance has been attached to the bureau's forecasts as a critical factor in state proration. If these estimates were discontinued, it seems quite likely that there would be no noticeable effect on state conservation practices and policies. This is not to suggest they do not perform a useful function as a check on a state's own estimates, or that the industry and analysts working on industry are not aided significantly by the forecasts.

Information on Stocks of Crude Oil

Of considerably more importance than the market-demand forecasts to state conservation agencies are the weekly reports on the levels of crude oil stocks. The Bureau of Mines released these reports through 1963, and, beginning in January, 1964, the API took over this function. These data are reported by companies on a voluntary basis and are broken down geographically by state and

---

[51] Second Report of the Attorney General on the Interstate Compact to Conserve Oil and Gas, op. cit., pp. 40–43, gives a discussion of the accuracy.

[52] See, for example, the grilling Alfred G. White got in his testimony before the Special Committee to Study Problems of American Small Business, op. cit.

[53] Second Report of the Attorney General on the Interstate Compact to Conserve Oil and Gas, op. cit., pp. 43–47.

by location, i.e., in tank farms or pipelines, at refineries. Stocks at refineries and in tank farms and pipelines are reported by refining and pipeline companies. Since there are relatively few companies involved, this reporting is quite complete. Crude stocks on leases are much more difficult to come by because of the tremendous number of leases and operators involved. For the nation as a whole, lease stocks account for less than 10 per cent of the total, refinery stocks account for from 20 to 25 per cent, and the remainder are in tank farms and pipelines. For Texas, the state that has roughly 35 to 40 per cent of the nation's stocks most of the time, crude in leases usually accounts for somewhat more than 10 per cent; crude at refineries accounts for only about 15 per cent; and crude in lines, tank farms, and terminals accounts for the rest.

State agencies use the weekly stock information to determine what demand pressures are currently at work on crude oil. A build-up of stocks anywhere along the line is an indication of supply exceeding demand, and clear evidence of this usually causes a commission to reduce production for the next month below the level that would have been allowed had stocks been "normal." Some states require the reporting of stock information so that the commissions are certain of having this information. In Texas, at least, the commission has in the past asked the major producers to indicate annually what they consider adequate stock levels to be in the United States, the United States east of the Rocky Mountains, and in Texas. This was done for crude oil and products separately. Currently, the Texas commission asks for adequate stock levels of the individual companies making nominations.

It is probably safe to say that state proration would be much less effective without adequate, up-to-date reporting of crude-oil stocks. Each commission must decide for itself what "normal" stock levels are; this is done by getting historical evidence and a "feel" for the situation. Variations from the normal seasonal swings in stocks are the best indicators a commission has of the need for more or less production.

Chapter 8

Conservation and the Market for Crude Oil

CONSERVATION AND OIL PRICES

One of the most difficult aspects of petroleum conservation to come to grips with is its relationship to crude-oil field prices. Critics of the present conservation regulations sometimes level the charge of "price-fixing" at the regulatory agencies or major crude-oil purchasers, or both. It is often also charged that the fixed price is unduly high. Supporters of the system of conservation regulation retort that beneficial "price stability," rather than price fixing, is achieved. They argue that the nature of the industry is such that chaos and waste result when there is not an orderly market and that a stable price promotes conservation. Some of these proponents agree that the present system of price support has its bad aspects, but the alternatives, they claim, are much worse. It is not our purpose to deal with the subject in this controversial vein, but rather to trace such relations as may be discerned between regulatory practices and market prices.

From our discussion in Chapter 1 and other places, it is quite evident that price is an integral part of conservation as viewed by economists. It will be recalled that we spoke of the economic view of conservation as one in which the discounted net-income (benefits) stream generated by our oil reserves is maximized over time. The key economic variables then are costs and prices (revenues), actual and expected. The extent to which conservation regulation affects costs and prices is, therefore, of great concern to economists, particularly if the effects in some way reduce the discounted net income (benefits) below what could be realized under some other set of circumstances. One great complaint that economists have of the present regulatory system is that its price and costs effects are glossed over.

In the public hearings held in the several market-demand states, prices are rarely, if ever, mentioned. Most regulatory agencies categorically deny that price is a consideration in setting state or well allowables. There is no reason to question this statement in ordinary circumstances or to suppose that the regulatory agencies are typically engaged in price manipulation. Nevertheless, it is impossible to speak of conservation without speaking of price; they are inextricably related. Certainly any general concept of conservation must include the thought that something of value is to be "used wisely," "saved," consumed or exploited more or less rapidly, more fully recovered, and so on. Prices, in a general way, reflect values that society collectively puts on various goods and services through a structure of markets. Prices also perform a rationing function in a relatively free economy such as ours, and conservation includes some concept of rationing.

By its very nature, conservation regulation entails some control over the market structure through which oil prices are determined. It is an error to suppose that regulatory agencies, individually or collectively, "fix" prices. What they essentially do is to specify some of the conditions under which buyers and sellers meet in the market, and by doing this they have a marked effect upon price terms. But it is necessary to know a good deal more about market organization and practices in order to see how this influence operates.

PRICE STABILIZATION AS A CONSERVATION GOAL

Our earlier treatment of the historical evolution of conservation controls[1] indicated that market stability was a major, if not the primary, goal of the efforts at regulation during the 1930's. Some advocates of this goal stressed the need for remunerative prices so as to assure a financially healthy industry. Others stressed the physical and economic waste of oil that was a consequence of chaotic market conditions. For whatever reasons, there was general agreement that higher and stable prices were a necessary condition for

---

[1] See Chapter 2.

successful conservation.[2] This was not a novel idea. The Oklahoma Legislature in 1914 had established a minimum price for oil taken from wells in the state.[3] The early arguments over possible federal controls, the Interstate Oil Compact, NRA controls, and state controls embodied the idea of market stabilization. In those earlier situations, stabilization meant not merely freedom from severe price fluctuations, but also stability at higher prices than those currently prevailing. The severe restrictions of output in states like Texas, Kansas, and Oklahoma in the 1930's were designed to have an immediate effect in changing the level of prices. This is not, however, true today; the present system *supports* prices, but does not actively attempt to revise them. It is necessary to distinguish between these earlier and later phases of the operation of the proration system. Since market stabilization is a major aspect of current conservation policy, we need to examine the mechanism with special reference to the price consequences.

The mechanism of market price stabilization is the proration system administered by the several oil producing states. This has been discussed in detail earlier, and we will not repeat that discussion here. In summary, this is a system in which estimates of current demand for (consumption of) oil are made and in which outputs are regulated so as to satisfy the estimated demand (consumption). The essential element missing from this sort of statement is price. Estimates are made of how much of the potential production of a state will find purchasers *at a price,* and the price is the current or recently prevailing price. This is the amount that is allocated to producers of a state. No estimate is made of "demand" in the sense in which that word is used in economic theory: a schedule of quantities that buyers will take at various prices. All that a state commission attempts to do is to estimate the "amount demanded" at a point on the demand curve. In other words, they take as a datum the current level and structure of oil prices, whatever they may be. It is this fact that permits regulatory officials to deny that they have any concern with prices, as illustrated by the following quotation from a prominent regulator:

---

[2] Stabilization can have several meanings. Generally in the oil industry it has meant minimizing the variance of prices around a trend, although never explicitly stated as such.

[3] This statute was never applied and thus has never been tested.

Question: Then would you care to express an opinion as to why the
            price of oil was increased in the last few weeks?
Answer: Why?
Question: How would you explain the increase in the price of fuel and
            oil in the last few months?
Answer: We have nothing to do with price. We are forbidden to con-
            sider economics; purely physical waste. I know nothing about
            price.[4]

From our own conversations with regulatory officials, we find it
to be true that this is, on the whole, the way they think about prices.
But those among them of a more analytical turn of mind understand perfectly well that the restriction of production is one of the
factors bearing directly upon the prices that are determined in the
market.[5] They are able to ignore prices because of two reasons:
(1) The demand for crude oil in the short run is highly inelastic,
and (2) through the market-stabilization mechanism, prices are no
longer a fundamental issue. A classic statement of the view of regulators about price is found in the 1964 IOCC study.

The argument is sometimes made that it is impossible to estimate or forecast market demand for crude oil except at a certain price. The answer
is that it has been done for years. The United States Bureau of Mines
and state regulatory agencies make such estimates or forecasts and do not
consider price. The state agencies consider many factors other than price
in order to determine the amount of crude oil that will be needed to
supply the demand for current consumption for the period under consideration.[6]

---

[4] Testimony of Ernest O. Thompson, member of the Texas Railroad Commis-
sion, at the hearings of the Committee on Interstate and Foreign Commerce,
House of Representatives, 85th Cong., 1st Sess. (1957), *Petroleum Survey*, p. 187.
The price increase referred to is the general crude-oil price increase that
occurred in the United States in January, 1957, after the closing of the Suez
Canal in the fall of 1956.

[5] "Although no state oil regulatory body gives consideration to price in determining market demand, it would be naïve not to recognize that market demand
proration does affect price. But the effect is to stabilize price and eliminate the
rapid and extreme fluctuations, which range from ruinously low prices to the
producer in times of oversupply, to disastrously high prices to the consumer in
times of scarcity. In addition to eliminating these extreme fluctuations, it results
in a higher average price to the producer and a lower average price to the
consumer." William J. Murray, member, Railroad Commission of Texas, "Market Demand Proration," in W. F. Lovejoy and I. J. Pikl (eds.), *Essays on Petroleum Conservation Regulation* (Dallas: Southern Methodist University, 1960),
pp. 72–73.

[6] IOCC Governors' Special Study Committee, *A Study of Conservation of Oil
and Gas in the United States* (Oklahoma City, 1964), p. 128.

It would have been much more enlightening if there had been some discussions of the impact of regulation on prices, and particularly an analysis of the behavior of regulatory agencies in situations where posted prices were raised or where fairly widespread discounts from posted prices were cropping up. We shall pursue some of these questions at a later point.

The mechanism, then, is one that sets production equal to estimated consumption (including necessary inventories) at existing price levels. Rather than a single body that makes the consumption estimate, it is a group of states that act independently from each other, each having considerable knowledge of the conditions in other states and the behavior patterns of other commissions. The statistical services of the federal government and the industry are essential elements in the accumulation and dissemination of this information. The result of this interdependent system is a great deal of over-all cohesiveness. The aspects of parallel (but not collusive) behavior on the part of the various state commissions, and the general unwillingness of any one state to seek aggressively a larger portion of the total national market for fear of upsetting its own local markets, are important in understanding the success of the stabilization policies.[7] The independence of each state is important to recognize also because it goes far in explaining why price setting is denied by each individual state—and the explanation is correct from the position that an individual state views the market and the industry. We find a strange paradox in which the state *individually* could disrupt market stability but in which any state individually could not achieve market stability. The states can, however, bring about stability if they act in an independent but parallel fashion.

The multi-unit decision making in state regulation has another important bearing on prices. Major crude-oil purchasers submit to each state commission estimates of the oil they will buy in the respective state. These are the purchaser "nominations" that are an integral part of the proration system. Many of these companies buy in several states, thus giving each firm a voice, albeit often small, in several state markets. Since state commissions maintain that they serve primarily to collect and evaluate consumption estimates of purchasers, the system's stabilization aspect may be more a function of company behavior than commission behavior. It does seem

---

[7] See Chapter 7, section 5, for additional discussion of this point.

true that commissions play more of a passive than an active role *in normal times* in the consumption estimating process. On the other hand, it also appears true that the purchasers acting without the commissions could not achieve the degree of stability that has existed in the regulated market without violating the antitrust laws. The structure of the crude producing and purchasing industry and its impact on prices and price behavior will be discussed below.

The preceding discussion might be summarized by saying that collectively the actions of state commissions in controlling supply have a stabilizing effect on crude-oil prices. They are not, however, price "fixers," either individually or collectively,[8] and their protestations against criticism that they do set prices seem in large part justified. Recent regulatory history reveals no attempts by any state or group of states to manipulate prices, in the sense of attempting to raise or lower them. On the other hand, their protestations that they do not have a great deal to do with determining the level of prices, particularly in times of overcapacity, are not justified. At such times, which now seem to be chronic, market-demand proration absorbs those pressures of excess supply that would normally cause prices to fall.

When production was fairly close to capacity in the early postwar years, prices moved upward during a period in which general prices were being adjusted upward. As overcapacity developed, it was prevented from having a depressive effect on prices by the market-demand proration process. The tendency of the system is thus to have a ratchet effect by which, when prices have risen, they are prevented from falling. This is a vast oversimplification, but perhaps gives a correct general notion.

The Suez situation was in a sense unique in this pricing process. With the closing of the Suez Canal in the fall of 1956, there was an increase in the demand for U.S. crude to supply the needs of Western Europe. There was, at the time, excess producing capacity in all the major market-demand states. Yet the posted price of crude rose about 25 to 35 cents per barrel. The producing capacity was not opened up by the state agencies and, as a result, they have been severely criticized. We shall not debate the feasibility of increasing

---

[8] After World War II, both Oklahoma and Kansas instituted minimum well-head price regulation for natural gas in some fields. This was ultimately discontinued because of the conflict with Federal Power Commission jurisdiction over producers' prices. *Cities Service Gas Co. v. State Corp. Commission*, 355 U.S. 391 (1958).

output at that time. Our point is that prices did rise 25 to 35 cents per barrel in early 1957 and then slid back about 10 to 15 cents per barrel over the next eighteen months or so under the pressures of overcapacity. Thus, we have a case in which the ratchet did not completely hold the initial price increases.

This collective behavior of commissions, be it conscious or unconscious, makes some contact with their fixation upon physical waste. The feeling is that ". . . prices that will yield a reasonable profit to producers perform a desirable function in stimulating drilling to replace production by finding new oil."[9] But at *any* price there are marginal wells and operators. If price falls, these marginal wells and operators fail, and in the eyes of many this is "physical waste," since some reserves are lost. It follows logically from this reasoning then that price declines are undesirable since they result in "avoidable physical waste."

THE CRUDE-OIL MARKET AND PRICING MECHANISM

The market for crude oil and the forces that determine crude-oil prices are complex and unique. We have noted earlier that much of the behavior in this market is parallel and interdependent, which is typical of oligopolistic market situations. However, the "orderliness" of the market is achieved primarily by way of state regulation rather than by the actions of the private participating firms. This concept must be further qualified to indicate that there is competition, albeit of a rather unusual type, and there are forces other than commission behavior making for price stability.

Crude oil is somewhat unique among raw-material markets in that crude-oil buyers make their purchases at the site of production. For many raw materials the producer moves the product to a market or central gathering point. But the crude-oil markets are in the fields, and the buyer usually bears the cost of transportation.[10] Few small producers can afford to provide their own transportation systems to refineries because of the small quantities of oil each has to sell, relative to the size of efficient pipelines. It is also true that the nature of pipeline operations makes it difficult and costly to segregate small quantities of crude flowing through a line so that

---

9 IOCC Governors' Special Study Committee, *op. cit.*, p. 92.
10 Since it is no doubt true that field prices would be higher if producers paid for transportation, the burden of this cost is probably shared.

the quantity, quality, or gravity of oil entering the line will be the same when it leaves the line. Large lease storage facilities that would enable producers to offer larger tenders for pipeline transportation tend to be quite costly, since full utilization of their capacity would be only intermittent. There seem, therefore, to be definite technical economies of scale that can be exploited by mass buying in the field and mass movement to refineries.

In 1964, about 84 per cent of refinery receipts of domestically produced oil was moved by pipelines, and probably 90 per cent of the crude moved by pipeline is handled by lines owned by relatively few companies that refine and market products. This contrasts sharply with the crude-producing industry in which there are thousands of companies, including most of the major companies that pipeline, refine, and market. The reasons for major-company ownership and control of pipelines are several, but probably most important is their desire to have adequate supplies of crude oil for their refinery needs. This is a critical need for a refining company, since refineries are characterized by sharply decreasing costs up to the capacity of an existing plant. Also, a refinery cannot be started and stopped intermittently without incurring substantial maintenance costs on the processing equipment.

While virtually all crude-oil trunk pipelines legally have common-carrier status, which means that they must transport oil for anyone who presents it to them, they are in practice much more like a private carrier or contract carrier. The reasons for this are obvious. The ultimate market for crude is the refinery. Pipelines are laid to serve a specific refinery or a group of refineries. Oil will move through a particular line if it is destined for a refinery and is unlikely to move through that line unless it is to be used in that refinery or group of refineries. There are several instances of non-integrated refineries located at the terminus of a pipeline, and such refineries can and do buy crude oil in the field and use an unaffiliated pipeline to move this crude to that refinery. In such cases, the pipeline cannot refuse to carry the crude. In 1965, the Chase Manhattan Bank study of the industry reported that the 29 "major" companies[11] accounted for about 88 per cent of the crude runs to

---

[11] There are several companies in the twenty-nine that only produce and do not transport or refine crude oil. Virtually all the major integrated refining companies are included.

stills, which indicates the predominance of the majors as buyers of crude oil from independent producers.[12] These same 29 majors accounted for about 56 per cent of 1965 domestic crude production.[13] In any specific regional market for crude, the role of the majors may be greater or less than this national average. It is also true in some instances that purchasers of crude may buy oil in the field and resell it to a nonintegrated refinery or to another integrated major company. This may be of some advantage to the nonintegrated refiner, since he avoids having to have a crude-oil-buying organization in the field.

If refineries with their affiliated pipelines are, in effect, the markets for crude oil in each field, is there really any competition on the buying side for crude oil? There is no complete answer to this that covers all situations. It is generally true that in Texas, Louisiana, Oklahoma, New Mexico, and Kansas the major fields are usually served by more than one line. In 1958, the five largest fields in each of these states were served by more than one line with the exception of two Louisiana fields that were offshore or in marsh areas. In a majority of these large fields there were three or more lines. In the East Texas Field there were 17 major lines and several smaller ones. In the Kelly-Snyder Field in West Texas there were five major lines. In the Sho-Vel-Tum Field in Oklahoma there were six lines. On the other hand, in Colorado only two of the five largest fields had two lines and the other three largest fields had only one line. In Wyoming four of the five fields had only one line with the fifth field having two lines.

This picture indicates more potential competition among buyers than is usually assumed by those discussing the nature of the crude-oil market. However, this "competition" among buyers occurs in the large fields and may be misleading when applied to all the fields in a state. For many, and perhaps most, small fields there is a single line and only one purchaser (usually an affiliate of the pipeline) to which to sell. In many instances the very nature of a pipeline precludes putting more than one line into a field. The field often just

---

[12] The runs to stills reported here include both domestic and foreign crude. If only domestic crude were considered, the percentage share of the majors would probably be somewhat less. *Financial Analysis of a Group of Petroleum Companies, 1965*, Chase Manhattan Bank (1966), p. 8.

[13] The Chase Manhattan Bank figures are for *gross* production, which includes all royalty oil.

does not have the producing capacity, or more correctly, the allowed production, to fill more than one line. To put two small lines into a field that could as easily be served by one large line is to ignore the substantial economies of scale present in pipeline construction and operation. An alternative means of transportation is by tank truck (or in the Gulf Coast marsh and offshore fields by barge). Trucks may be the lowest-cost form of transportation for given small or remote fields. Trucks may also be used where pipeline capacity is insufficient.

In the longer run, for any field served by a pipeline there is also the potential competition from a pipeline that could be built to serve that field. This threat is probably not too significant since a new line to a field that cannot produce enough to fill two lines can only result in losses for one or both lines. In some fields, however, it no doubt does serve as a real threat.

The physical arrangement in an oil field served by two or more lines involves actual pipe connections between the lease tanks and pipeline. Such connections are not costless and frequent shifts in connections by producers from one line to another are part of the cost of connection. Thus both buyer and seller have an interest in maintaining the connections once they are made. This may mean that a pipeline finds it in its own self-interest to satisfy its connected producers if possible. This means, in many situations, that short-run surpluses are usually insufficient grounds for a purchaser to disconnect and seek cheaper crude elsewhere, and short-run shortages are insufficient to cause producers to disconnect to seek a higher price. Both buyer and seller have an important interest in the continuity of the market. This goes part of the way in explaining why price discrimination, which would appear to be a likely phenomenon in such market situations, is not often practiced by purchasers. Other deterrents to price discrimination will be noted later.

Crude oil is usually purchased under a contract known as a division order in which the purchaser is authorized to divide the proceeds of the sale among the various property interests in the production. The division order does not specify a price but merely refers to the price that is posted by the buyer in that field for a specific gravity and quality of oil. This contract usually can be terminated by either the buyer or seller upon written notice. In some instances, oil purchased under a division order is sold immediately to another purchaser at the price paid. The first purchaser makes the division of revenue and retains his right of purchase. The second purchaser may have a refinery nearby that he wishes to supply. Often such a transaction will include an agreement by the second purchaser to sell a quantity of oil to the first purchaser at another location. This gets into the rather complex area of exchange agreements and crude-oil trading among purchasers.

Exchange agreements are a system by which several companies, each owning production or having purchased crude scattered geographically, can sell or exchange with other companies so that each minimizes its transportation costs on the oil needed to supply its refineries. The system also enables each purchaser to obtain or dispose of more easily the quality or type of crude he needs or does not need in the desired volumes at the time he wants to. Exchanges may also involve products traded for crude. It can be argued that this type of system assists a producer in that he is not dependent upon a particular purchaser's needs for the specific type of crude being produced. It clearly gives refiners more certainty of crude supplies of desired types and thus gives the refiner flexibility. This increased certainty may reduce the need for crude-oil storage capacity at refineries.

Posted pricing for crude oil in the field is a variation of pricing systems that are often found in oligopsonistic markets in which a few buyers purchase from many sellers. It is, in effect, an "offering price" by the purchaser, similar to the situations found in many commodity markets. It differs in that the buyer-seller relationship usually is long term, i.e., it lasts for more than one transaction. The reasons for the long-term nature of the relationship were presented in the preceding discussion.

The effects of a posted pricing system have never been thoroughly analyzed, and we can only point out what have been advanced as its advantages and disadvantages. One obvious effect is to give considerable stability to crude prices. It is perhaps arguable whether clear-cut rigidity is introduced through this mechanism. With prices being posted in every field, and often by several purchasers in a single field, both buyers and sellers have considerable knowledge of the market—more knowledge than in many markets. This probably acts as a deterrent to price discrimination by purchasers. It allows a seller to view what other sellers are receiving for similar crude similarly located with respect to markets. Purchasers must post prices that assure them of long-run supplies of oil.

The price posted by a particular purchaser does not mean that other purchasers must also purchase at that price. It means that the purchaser posting the price will execute the division orders he has in effect in that field at that price. Other buyers may post different prices in a given field, and purchasers who do not post prices may buy at other than the posted prices. Such situations are not common, but do appear to occur with enough frequency to lend support to the argument that there is not complete price rigidity in this market. It is conceivable that different postings for the same crude in the same field may reflect differences in transportation costs for different purchasers to their refineries or terminals. This becomes a problem in logistics in which each purchaser is attempting to minimize his costs of crude oil *at his refinery* and not in the fields. Higher posted prices by a purchaser in some fields may be justified if they attract the desired types and quantities of crude in locations that have the lowest moving costs to his refinery.

There is, in this system, the complicating fact that many purchasing companies also produce oil. In a field in which a purchaser owns substantial production (but not all of it), there may be an incentive to post a high price. The incentive comes from a judgment on the part of the company to take more profits at the producing level and less at the manufacturing or refining level. Crude-oil inputs to a refinery are a cost, and the owned production in this input is charged as a cost of refining in an interdepartmental book transaction. A high cost of self-produced crude means a high refining cost, but it may also mean a high profit in production. A dollar of before-tax profit in production yields a greater after-tax profit than a dollar of before-tax profit in refining. Such a statement is a considerable oversimplification but seems to be generally true. The reason for this is the presence of special federal income-tax treatment of expenses in and revenues from drilling and production. Whether this is a motivating force in posting higher prices in some fields by some purchasers, we do not know. It would seem, however, to be a logical profit-maximizing behavior for an integrated company.[14]

---

[14] For a discussion of this point, see M. G. de Chazeau and A. E. Kahn, *Integration and Competition in the Petroleum Industry* (New Haven: Yale University Press, 1959), pp. 221–29.

One of the most important effects of the posted pricing system is that it facilitates crude-oil exchange among major purchasers. It gives detailed knowledge of prices all over the country and thus provides a fairly simple basis upon which to calculate the relative values of crude oil being exchanged. It is impossible to know to what extent crude is exchanged at posted prices or something other than posted prices. While there might be reasons for a given major company to hold up another major company for a higher price, the conditions making this possible are apt to be short lived. The ill-will created might not be wise long-term policy, since in the future the shoe may be on the other foot. It should also be noted that the systems of exchanges and posted pricing interconnect the thousands of local markets in the fields and create a national market. Obviously, there are not uniform prices for crude throughout the country, even for like gravities and qualities. It is true, however, that prices in all fields are interrelated, with each affected by all others. This helps to explain why a price change in a given field may eventually result in price changes in the same direction in virtually all fields. This, in turn, may help to explain the stability of oil prices generally.

While it is true that price stability characterizes crude-oil markets, it is also true that some crude does not move at these posted prices. Premiums and discounts may be used to attract or discourage supplies, depending on the market conditions. If a particular purchaser finds that he is in short supply of a particular quality of crude, a fairly small premium may be sufficient to induce sellers to shift if gathering systems make it physically possible. Often, such a situation may be regarded as a short-run phenomenon by a purchaser, and he will deem it wiser to give a premium that he can drop relatively easily, rather than to raise the posted price, a move which he may find much more difficult to reverse. If discounts or premiums become widespread in an area, it is likely that they will be formalized by a change in the posted price. This behavior is also fairly characteristic of oligopsonistic industries.

Unfortunately, there are no recent published data showing premiums and discounts that will give us a clearer picture of short-run market forces for crude oil. It is much more likely that nonposting purchasers will use premiums and discounts than posting purchasers. If a purchaser takes a discount from or adds a premium to his posted price and does not make this uniform among sellers in a field, he is practicing a form of price discrimination. Even if this is not illegal (and it may be, depending on the state of the situation), it is apt to lose him considerable goodwill among sellers. If the discount or premium is uniform among sellers, it may reflect the situation described above in which there is a short-term surplus or shortage, but one not serious enough to justify a change in the posted price.

It is extremely difficult to get a complete grasp of the market forces affecting crude-oil prices. There are, however, numerous straws in the wind in the form of comments or stories on prices in specific fields or regions. Between 1962 and 1965 there was a gradual deterioration in crude-oil prices. This was reflected in decreases in posted prices by some purchasers and by discounts in some instances. For example, between January 1 and October 15, 1963, the *Oil and Gas Journal* reported that posted price cuts had affected over 1,200,-000 barrels of daily production, with a resultant loss of more than $11 million to producers.[15] The largest losses were reported to be in Oklahoma, Kansas, and Southwest Texas. The same journal reported in June, 1964, that producers' income had been trimmed more than $30 million during the preceding seventeen-month period.[16] The cuts in late 1963 and 1964 were most severe in Kansas, the Rocky Mountain fields, and the Illinois–Ohio region. This rather gradual deterioration continued into 1965 despite continued severe restrictions on production in market-demand states. By late 1965, the market situation had firmed up considerably in many areas. By the spring of 1966, a number of major companies purchasing substantial quantities of crude oil raised posted prices 5 to 10 cents per barrel. In May, 1966, the IPAA is reported to have indicated that prices had risen an average of 7 cents per barrel on about 720,000 barrels of the nation's 8 million barrels of daily production.[17] Additional small increases were reported in June and July, 1966, for other production.[18] The spring of 1966, as we noted earlier, was also a period of rising allowables.

---

[15] October 21, 1963, p. 55.

[16] *Oil and Gas Journal,* June 8, 1964, p. 93.

[17] *World Oil,* May 1966, p. 13.

[18] *Oil and Gas Journal,* May 16, 1966, p. 160; June 6, 1966, p. 62; July 11, 1966, p. 56; July 16, 1966, p. 54.

It was quite evident that the regulatory commissions noted the price erosion in the 1962–65 period, but the manner in which it occurred made it difficult to control. There were very few general price cuts in any one state. If a cut were announced in west central Texas, for example, it might very well reflect oversupply in that one part of the state. But the Texas regulation (and those of other states also) are not well designed to handle a regional problem that is not general within the state. Thus, while some producers in the north and west central areas of the state in the spring of 1965 proposed a complete 15-day moratorium on production in the entire state, the Railroad Commission did very little except urge purchasers to find some way to take the "distress" crude.[19] Under such circumstances a commission can do little, yet such isolated price deterioration over fairly long periods can ultimately reduce prices generally.[20] The same is true for price increases of the sort that occurred in 1966. Shortages of particular types of crude in specific regional markets are reflected in local price increases. State regulations do not permit adjustments in local supply situations to permit more production, even though unused capacity may exist in the particular area in which a price increase occurs.

State officials and independent producer groups were not alone in their concern about the sliding crude prices. In an unprecedented statement, the Secretary of the Interior noted: "The Department of the Interior supports a vigorously competitive petroleum industry. The economic strength of the major factors in the industry, however, should not be utilized as an unfair competitive weapon in dealing with small nonintegrated crude producers." Noting that product prices in the first half of 1965 were quite firm, he stated that, "In these circumstances, and particularly in light of the earnings position of the major firms in the petroleum industry, care should be taken in crude pricing practices to avoid actions which could be contrary to the security interest of the United States in maintaining adequate petroleum reserves." In noting that the im-

---

[19] In June, 1965, a group of Texas independent producers chartered a corporation to be a nonprofit purchaser and reseller of crude. This is an attempt to find a market for crude in some of the distress areas of the state.

[20] One member of the 1963 Texas Legislature proposed legislation that would give the Railroad Commission the authority to set field prices for crude. This legislation was not passed.

port control program and state conservation created stability in crude supplies, he concluded that "all the ingredients necessary for crude price stability are present."[21]

The implications of Secretary Udall's statement remain unclear. It is unusual, since, as far as we are aware, no federal agency has explicitly expressed concern over price softness in crude oil since the days of the NRA in the early 1930's, though federal officials have on occasion praised the states for maintaining market stability. The few cases of recent federal intervention in industry pricing generally have been on the side of opposition to price increases, as in the case of President Kennedy's "rollback" of steel prices in 1962 and two or three similar interventions by President Johnson on anti-inflationary grounds.

It may be that Secretary Udall was picking up a point made by the Attorney General in his 1962 report on operations under the Interstate Oil Compact.[22] In this report, the Attorney General dwelt at some length on the strategic advantage held by the major-company owners of the pipelines network in relation to scattered independent producers. He suggested the existence of abuses of power that might call not only for antitrust action, but possibly new legislative action. The wording of Secretary Udall's statement— "The economic strength of the major factors in the industry . . . should not be used as an unfair competition weapon in dealing with nonintegrated crude producers"—suggests that he may be attributing the price softness to an abuse of competitive power by large purchasers.

The reaction in the industry to the statement was mixed, with the predominant feeling apparently being that the fixing of prices was no concern of the government. Some producer groups suffering from price cuts applauded the statement. It seems highly unlikely that the federal government can do very much about the price situation in oil. It could tighten imports, but this at the moment appears unacceptable to it and, in any case, imports appear to have had little to do with recent price changes for crude oil. Antitrust investigations could be initiated, but this is a devious and time-consuming

---

[21] U.S. Department of the Interior, Office of the Secretary, press release, June 9, 1965.

[22] U.S. Department of Justice, *Report of the Attorney General Pursuant to Section 2 of the Joint Resolution of August 7, 1959 Consenting to an Interstate Compact to Conserve Oil and Gas* (Washington, 1962).

approach which would have uncertain results and which the facts might not warrant. At present the facts underlying recent price behavior are extremely obscure.

If the federal government's concern is with the financial incentives for producers to find and develop new reserves, it would appear that the obvious has been overlooked. Incentives can be increased by reducing costs, and costs could be reduced by more attention to revisions of state conservation regulations. To the best of our knowledge, the Interior Department has made little effort in the direction of studying industry costs and suggesting ways in which these could be reduced. This, however, leads to another line of discussion that we shall not pursue at this point.

There has occurred in recent years a somewhat perplexing phenomenon in crude-oil purchasing that may possibly have had some peripheral relation to price behavior in some regions. Several major integrated oil companies that formerly maintained extensive gathering and trunk pipeline systems and that purchase substantial quantities of crude oil have sold pieces of their pipeline systems and, in particular, their gathering systems. The reasons for these moves are not entirely clear, but it is possible to report some speculations about them. Some major companies have been actively acquiring, through purchase and merger, numerous nonintegrated producing companies. This has meant that more of their refinery inputs come from their own producing properties, and they have a consequent lesser need for purchased crude. The state laws that require ratable take and prohibit discriminatory purchases among fields would make it difficult under an allowable system for major company-owned gathering systems to reduce purchases of non-company-owned crude. However, the selling off of some gathering systems in those areas in which company-owned production was small or nonexistent relieves the integrated company from purchasing that oil. The independent gatherer now has the burden of taking oil ratably and in a nondiscriminatory fashion, but the refiner–purchaser does not. The localized gathering system may be able to put downward pressure on crude-oil prices in the particular area or to pass on some transportation costs to the producer. If this pressure actually results in lower prices in selected areas, the major refiner–buyer may benefit to the extent he purchases this crude from the gatherer and to the extent the price decrease is passed along to him. If this occurs, it could be a very profitable arrangement for the gatherer. This is a problem area that needs investigation but that takes us away from our central topic of conservation.[23]

The market for crude oil, therefore, is one in which there is substantial oligopsony, or even monopsony in individual fields, but also one in which local markets are interrelated. There is a relatively stable general level of prices with small changes within the price structure occurring fairly frequently; changes or shifts in the whole structure are rare. This price stability reflects in part the small number (perhaps twenty) of major crude-oil purchasers in the national market. The potential initiator of a price change is hesitant about making a change because he knows that he is large enough to influence the general level of crude prices, and he knows his competitors will react, but he does not know for sure how they will react. A price increase initiated by one purchaser but not followed by others puts that purchaser at a disadvantage in refining and marketing. A price decrease initiated by one purchaser may elicit equal or greater price decreases by other purchasers and will certainly arouse ill feelings among sellers. In addition, a price decrease may adversely affect the profits from an integrated producer's own production.

Buyer–seller relationships seem strongly influenced by long-run considerations of stable supplies for buyers and stable outlets for sellers. Knowledge of market conditions by both buyers and sellers, while not perfect, is extremely good.

Superimposed on this rather peculiar free-market mechanism is the system of state control of supplies. Such a system clearly tends to act as a buffer against free-market forces that would perhaps otherwise make for change. The effect is to dampen price movements. The state control system is fragmented so that there is no single body controlling supply, but rather a group of bodies that do not appear to work cooperatively in this function in any explicit sense. Finally, there is the interrelationship between the free-market forces and the state controls. Estimates of consumption by state authorities are determined in part from purchaser nominations. Thus, in an indirect way the purchasers determine both supplies and demands in the several states in which there is market-demand proration. In states where there is no market-demand proration,

---

[23] The 1965 Texas Legislature amended the Common Purchaser Act in Texas (which requires ratable taking and no discrimination) so that it covers independent gatherers.

demand (at the going price) is reflected in what purchasers take in the fields. There is, however, no evidence to our knowledge of collusion or cooperation among purchasers in fixing prices.

PRICES AND THE NATURE OF
DEMAND FOR PETROLEUM

The demand for crude oil, as is true for most raw materials, is a derived demand. It depends upon the demand for the refined petroleum products: gasoline, fuel oils, kerosene, lubricating oils and greases, asphalt, road oil, waxes, and petroleum-based chemicals and their thousands of derivative end products. Crude oil differs from many other raw materials in that the end products are so numerous and serve so many different markets. We shall not attempt to do much in the way of analysis in these markets, but a few observations about the markets for gasoline and fuel oils will shed some light on the demand for crude oil itself.

In the United States the bread-and-butter product for most refineries is gasoline, with distillate and residual fuel oils being second and third in importance. The average barrel refined in the United States in 1963 yielded 44.1 per cent gasoline, 23.9 per cent distillate fuel oil, 8.6 per cent residual fuel oil, 5.1 per cent kerosene, 3.5 per cent asphalt, 3.1 per cent jet fuel, and the remainder in miscellaneous products.[24] A breakdown of industry revenues by product would give gasoline an even larger share and residual fuel oil a smaller share.

For the industry (as contrasted with a firm in the industry) it appears that the demand for gasoline in the short run is relatively inelastic in the range of current prices. The published and unpublished work on this subject seems to bear out this conclusion, as does common sense.[25] A decline of 5 cents per gallon for gasoline would probably not cause the amount of gasoline consumed to increase substantially except in the very short run during which time consumers might fill up the tanks of their automobiles. Such a price drop, about 17 per cent of the retail price of 30 or 31 cents per

[24] U.S. Bureau of Mines, *Minerals Yearbook, 1963, Vol. II: Fuels* (Washington: U.S. Government Printing Office, 1964), p. 448.

[25] See Ralph Cassady, Jr. and Wylie L. Jones, *The Nature of Competition in Gasoline Distribution at the Retail Level* (Berkeley: University of California Press, 1951).

gallon (including taxes), would not induce consumers to drive their automobiles much more. Nor would a 5-cent-per-gallon increase in price reduce gasoline consumption significantly. Over longer periods of time it seems likely that the price elasticity for gasoline is greater, though we can cite no evidence.

In the United States, at least, gasoline appears to be relatively income inelastic also in the short run, although published evidence on this is scarce. Over long periods of rising family income there is a tendency for the ratio of automobiles to families to rise, with a consequent increase in gasoline consumption per family, but this probably has a tendency to reduce gasoline consumption per vehicle. Also, with rising income there may be a tendency for families to purchase larger cars having greater consumption per mile. These are rather long-run phenomena, however, and do not seem to have any identifiable price effects.

Recently the API and other industry trade associations have launched an advertising campaign to induce people to drive more, particularly for vacations. It is too early to tell whether these efforts have been successful, but the likelihood of significant gains in gasoline consumption seems remote.

The demand curve for gasoline facing any individual refining–marketing firm in the industry is probably quite elastic in the relevant price range. Any increase in price at the retail level by one firm will cause consumers to switch in large numbers to another brand of gasoline. A reduction in price by a single firm will result in significantly higher consumption of this firm's products, if other firms do not follow the price cut. However, price cuts are usually met fully and quickly by other firms, thereby maintaining about the same market division as before. This characteristic of the market, in effect, presents the firm with a "kinked" demand curve, one of the classic models described in oligopoly theory. There develops in this situation a "live and let live" pricing policy in the industry in which there is considerable conformity to established pricing patterns, disturbed at times in local markets by companies attempting to expand their market shares. These patterns include accepted differentials between major brands of gasoline and minor brands sold by "independent" retailers. Competition is then of a nonprice variety with service, sales techniques, and alleged product differentiation used as the primary devices for increasing an individual company's share of the market.

Comments about the fuel-oil markets cannot be generalized as readily as for gasoline. Distillate fuel oils go primarily into space heating and railroad locomotive uses. Residual fuel oils are used principally for industrial heat and power, space heating, bunkering of vessels, and railroad locomotives. Some of these markets face vigorous competition on an industry-wide basis from substitute fuels such as natural and bottled gas for space heating, and natural gas and coal for primary industrial heat and power. The result is that these fuel oils are priced to compete with other fuels and the demand curves for them facing the industry are apparently fairly elastic in the relevant price ranges for all but short-run situations. For individual firms the demand curves are quite elastic, reflecting the presence of competing firms as well as substitute fuels. In addition, about one-half of the U.S. supply of residual fuel oil is imported primarily from Venezuela and the Netherlands West Indies. This is low-cost fuel oil and tends to put a ceiling on prices for domestically produced residual oil.

It may be concluded then that the products of petroleum face a variety of market situations. U.S. refineries can vary the product mix of their output, but tend to produce high gasoline yields, particularly in spring and summer months, and fairly high distillate heating-oil yields, particularly in fall and winter months, and minimize residual fuel-oil yields. The demand of refineries for crude oil tends to move with secular growth, lagging somewhat, however, because of the rapid growth of demand for natural gas in competitive uses. Swings in product demands that usually reflect variation in weather conditions and to a lesser extent fluctuations in industrial activity do not cause swings in the demand for crude in as great an amplitude because of the ability of refineries to change the product mix. For example, severe winter weather may dictate higher heating-oil yields, but gasoline inventories may not decline greatly because inclement winter weather reduces winter driving somewhat.

COST CONSIDERATIONS AND CRUDE-OIL PRICING

Behind the supply of any commodity lie the cost characteristics of producing the commodity. For domestically produced crude oil, costs are said to be "high" relative to the costs of producing in South

America, the Middle East, and North Africa. Evidence to document this claim falls far short of accurately quantifying a representative cost of finding, developing, and producing a barrel of oil, but what evidence there is corroborates this conclusion.[26] Since it is the long-run marginal cost of oil at points of consumption rather than average costs at points of production that are relevant in determining the amounts that areas of the world will produce, comparisons of finding, developing, and producing costs among countries can be quite misleading. Even if the world market for crude oil were completely free of all restrictions, the Middle East and North Africa areas, which are reputed to have average costs at exporting points of less than 25 cents per barrel, would not serve all the markets of the world. It is the long-run cost of the incremental barrel at the point of consumption, including long-run finding and developing costs, that dictates what areas will serve what markets. Domestic production no doubt generally has higher long-run marginal costs than much foreign production even in U.S. markets. The relevant point to note here, however, is that U.S. costs are unnecessarily high and could be lowered. Any lowering of long-run marginal cost would permit a lowering of the price and thus increase the share of U.S. production in the domestic market, in the absence of import controls. From the standpoint of increasing economic efficiency, conservation regulation could do more than it has done toward reducing the costs, and much of the burden of this study has been to point out areas of conservation regulation in which cost savings could be realized.

From the nature of the occurrence of oil and the technology and structure of the industry, it is clear that some oil is high-cost while other oil is relatively low-cost. It can be assumed that current domestic oil prices are adequate to cover at least the short-run marginal costs of existing wells at current rates of production. This is an obvious price floor. It is conceivable that total costs are not being covered for some wells, but the sunk costs are irretrievable, and it is worthwhile to produce a well as long as something more than out-of-pocket costs are being recovered. It is, therefore, impossible to say how much oil or how many wells are not paying their full

---

[26] See W. F. Lovejoy and P. T. Homan, *Cost Analysis in the Petroleum Industry* (Dallas: Southern Methodist University Press, 1964), for a discussion of major postwar cost studies.

costs at current prices.[27] If substantial quantities of oil are in this situation and expectations are that much future oil may be in a similar situation, a natural result would be a diminution in drilling, especially exploratory drilling but probably developmental drilling also. There might also be a diminution in secondary-recovery efforts, although the forces operating here are perhaps more complicated, since this might be a cheaper way to acquire additional reserves than new drilling for some companies, and proration treatment of secondary-recovery projects is often different. Second, there conceivably would be pressure on prices to rise.

Trends in drilling activity were presented in the discussion of excess capacity,[28] and need not be repeated here. It is sufficient to note that while drilling activity has declined from the peak levels of the late 1950's, it is not clear whether this decrease has at this point been sufficient to begin the long process of permanently reducing excess capacity. Continued high levels of drilling activity would seem to indicate that for many individual operators, at least *expected* revenues will be sufficient to cover *expected* costs and return them a reasonable profit. Since revenues are determined by multiplying price times quantity, the continuation of fairly high drilling rates must mean either that operators (1) find current prices and allowables are adequate, or (2) expect future prices will rise and, even at current or lower allowables, will provide sufficient revenues, or (3) expect future allowables to rise so that, even at current or lower prices, revenues will be sufficient, or (4) expect both prices and allowables to increase to provide adequate returns. If we had to guess, we would say that domestic crude-oil price increases seem highly unlikely any time soon, given existing excess producing capacity, relatively low-cost foreign crude supplies, and the presence of shale oil and tar-sands oil whose costs are approaching the competitive level with conventional crude oil. It remains to be seen whether the firming of prices in 1966 will be permanent. We would also guess that allowables will not be raised significantly in the near future, although the possibilities may differ from state to state.

How cost factors may affect the industry supply situation in the

---

[27] For an interesting discussion of the application of economic theory to petroleum costs and prices, see M. A. Adelman, "World Oil Outlook," in Marion Clawson (ed.), *Natural Resources and International Development* (Baltimore: The Johns Hopkins Press for Resources for the Future, Inc., 1964).

[28] See Chapter 5.

face of fairly firm ceilings on prices and allowable rates of production is highly conjectural. The subject of cost-supply relations for crude oil is intricate, and leads head-on into fallacious reasoning unless developed with care and at length. Furthermore, an adequate study of costs would require an array of data on the producing capacities and costs of existing wells. These data do not exist except in a completely unorganized fashion perhaps in the hearings of the various state commissions on specific fields. Since we cannot take the present occasion to pursue the subject, we will make only one or two observations in the context of conservation regulation. Substantial changes in regulations that would reduce the operating costs of *existing* wells are quite conceivable and desirable, but are unlikely to be made., Working against such changes are powerful political groups that have a vested interest in no change. These interests find strong support in the courts, which are reluctant to impinge on property rights for the sake of greater efficiency. Any changes in regulatory practice that can lower the *future* costs of finding and developing new oil will affect aggregate industry costs only gradually, since the wastes and inefficiencies of past practices are imbedded in existing reserves and producing capacity. In the degree, however, that cost-saving is achieved, it should add to the incentives for exploration. This could lead in turn into a circular puzzle: that greater incentives may cause greater productive capacity, which leads to greater production restrictions, which causes unit costs to rise, which tends to reduce incentives. We do not wish to predict that this dead-end situation will necessarily appear, but the circular form of statement suggests certain relationships of great potential importance for regulatory practice. Something more will be said about this in the concluding chapter.

Chapter 9

A Summary Review

The existing structure of the crude-oil-producing industry in the United States is, in large degree, a result of the programs of state regulation that have been followed for the past generation. The primary purpose of the present study was to examine the consequences of those programs from a particular point of view—that of applying to them certain tests to determine how well they conform to commonly accepted principles of economic efficiency. Underlying this purpose lay the presumption that efficient use of oil resources is a matter of public interest, including the oil industry's interest, and that deviations from those principles require special justification. As the study proceeded, it became increasingly evident that, while conservation regulation had accomplished much in eliminating waste in the development and use of oil resources, the pursuit of economic efficiency had been significantly impeded by a popular philosophy of property rights and individual enterprise, by inertia created in large part by conflicting vested interests within the industry, and by a substantial lag on the part of state legislatures and regulatory authorities in recognizing major changes in technological and economic conditions. It became further evident that the regulatory rules and processes had brought the industry face to face with certain problems bearing upon its future development.

BACKGROUND

Looking backward at the history of conservation regulation, the most astonishing fact to an outside observer is that the interests desiring to make money out of producing oil have supported policies that required them to incur much greater costs than were neces-

261

sary and that greatly curtailed the amount of oil they could ultimately recover from their properties. This does not look at all like
the behavior of "economic men" who are supposed to be eager to
maximize their profits. We have traced the source of this noneconomic behavior to two things. First, there is a philosophy of
private property rights and individual freedom of action that is
deeply imbedded in the American tradition. To have grasped the
opportunities for economical development of oil resources would
have required a degree of cooperative action and a relinquishment
of individual discretion quite foreign to habitual patterns of behavior. Second, the individual operator has no choice but to look
at the rules of the game and formulate his profit-maximizing behavior within the framework established by these rules. If the rules
do not allow efficient operations from an over-all industry viewpoint, then the individual operator can only behave efficiently and
profitably within the prescribed boundaries. This leads back into
the first point. The operators collectively can influence the establishment of efficient rules, but this takes cooperation that has been at
times impossible to achieve because of conflicting entrenched economic interests within the industry.

The source of the peculiar discrepancy between private economic
interest and the actual processes of development may be traced to
the origins of the industry. Private surface owners took advantage
of their right of access to the underlying oil without regard to the
necessity of cooperative action in order to develop reservoirs with
full economic effectiveness. The early results were so intolerably
wasteful in the physical sense and so disastrously disorderly for the
market situation that regulation was imposed to create some order
in the prevailing chaos. The types of regulation imposed were, however, never primarily designed to achieve economic efficiency in the
sense of minimum costs, but only to modify the forms and degrees
of individual enterprise in the production and marketing of oil.
Waste prevention seemed often to be incidental to the maintenance
of the established system of private property rights. The desire of
owners and operators to retain a large degree of individual discretion in the development of their properties was, and has continued
to be, the primary source of the extravagantly wasteful organization
of production.

An explicit statement of the dual purpose of regulation may be
found in the 1964 report of the IOCC Governors' Committee:

The economy of each state in which the production of petroleum is significant is improved because of the existence of the petroleum industry in the state. . . .

It necessarily follows that the prevention of reasonably avoidable waste or loss of petroleum, a non-renewable resource so vital to our lives, is essential. The Public interest is evident. . . .

One of the important functions of a state is to protect owners of land in that ownership, and to resolve conflicts of rights, and to enforce duties incident to ownership and the enjoyment of ownership. Indeed, conservation statutes of the states, while justified as being in the public interest, are also justified as being necessary to resolve conflicts incident to private ownership of land.[1]

This passage states the central conflict of purpose in the regulatory process: between avoidance of waste on the one hand and concern for various kinds of property interests on the other—the two being in certain respects irreconcilable.

We should stress, as we have tried to do from time to time in the text, that oil-conservation regulation has undoubtedly made a significant contribution to oil conservation in the economic sense when compared with what would most likely have happened in the absence of any sort of regulation. We should also like to note that there have been significant advances in the direction of what we call economic efficiency, particularly since 1950. From our economic view we feel, for example, that the moves toward some unitization —both compulsory and voluntary—is better than no unitization, that the "new pool" Louisiana Yardstick and the 1965 Texas Yardstick are better than their predecessor yardsticks, that almost compulsory 40-acre minimum spacing now found in Texas and Louisiana is better than 20-acre spacing, that compulsory pooling is better than the absence of compulsory pooling, that maximum allowed gas-oil ratios are better than no requirements on such ratios. This list could be extended. Our analysis suggests additional, more rapid, and in some cases, from the industry's view, more radical changes in regulation. We are not advocating the destruction of the regulatory system but rather what we hope are constructive suggestions that will improve the economic health and prolong the useful life of the domestic oil-producing industry, and also that will concurrently serve the public interest.

---

[1] IOCC Governors' Special Study Committee, *A Study of Conservation of Oil and Gas in the United States* (Oklahoma City, 1964), p. 11.

As matters now stand, an extremely complex production and market structure has been built upon the basis of regulatory ground rules which, in many respects, do not conform to the requirements of economic efficiency. There is, however, no easy way to achieve such conformity. An industrial structure is an organic thing, not something that can be taken apart and reassembled to the heart's desire. Policy starts from where you are and is limited by what can be done to modify, without destroying, the organism. When dealing with what Professor John R. Commons called "going concerns," the procedure is to identify the nature of a problem and to devise the practicable means to some definable ends. There is a tremendous inertia in the regulatory system based on statutory law, court decisions, established rules, contractual relations, vested interests, and habitual behavior. Established rules and habits are unlikely to undergo substantial changes unless subjected to internal stresses or external pressures. The literature of the industry has recently been full of evidence that such stresses and pressures are gaining strength and that change is in the air. A statement by the API, noting a ground swell of discontent, states:

To meet growing future requirements (for oil and gas), continued progress in technology and operating practices will be needed and substantial improvements in conservation practices and regulations will be imperative.[2]

The Secretary of the Interior in 1963 addressed to the Interstate Oil Compact Commission a statement of his concern over unsolved problems in conservation, including the following passage:

Several factors have worked together to limit the effectiveness of their control system. Among these are the status of state regulatory statutes in light of present day technology, the limited participation of some producing states, and the increased flexibility of interstate purchasing and transportation facilities.[3]

Individual regulatory authorities, industry representatives, and outside observers have issued numerous statements attacking what they consider to be defects in the existing system and advocating constructive changes.

---

[2] API, *Statement of Policy: Conservation, Development and Production Practices* (New York, 1963), p. 3.

[3] U.S. Department of the Interior, press release relating to letter of April 4, 1963 from the Secretary to the Chairman of the IOCC.

While there is no consensus among affected interests as to precisely what changes are called for, the deep sources of trouble are, we believe, those which have been outlined in this study. That is to say, they are centered in the economic factors that generate excessive costs and unused capacity. The influence most likely to activate the industry to a demand for change is a narrowing of the margin between costs and revenues. But new elements of the public interest also appear likely to affect the outcome. State governments are torn, on the one hand, between the politically potent desire to protect the vested interests of groups and communities dependent on the existing structures of the industry and, on the other hand, the need to promote the full future development of their oil resources. The federal government is deeply involved in a concern for the adequate future availability of energy. It might also conceivably develop a new interest in consumers' desires for low-cost oil. While it is a matter of sacred doctrine in the states that the federal government should stand clear of the processes of state regulation, the federal government is in a strong position to bring pressure to bear because of the three vital ways in which it is supporting the industry: (1) through import restriction; (2) through favorable tax treatment; and (3) through acceptance of state regulations on federally administered lands.

If the federal government develops a more active interest in the cost and adequacy of domestic energy supplies and urges, as it has, the discovery and development of additional oil reserves, the state governments and the industry are faced with the problem of how to maintain or increase reserves while keeping to a minimum the additions to producing capacity so long as an excess remains. We submit that additions to reserves will continue to lag as long as economic inefficiencies built into state conservation regulations remain unchanged.

SOURCES OF ECONOMIC INEFFICIENCY

In the course of our study we have identified four interrelated, but separable, sources that define the nature of the industry's inefficiency and underlie its basic problems. These are (1) excessive capital investment, (2) excessive producing capacity, (3) inefficient reservoir development, and (4) proration rules that favor the production of high-cost oil. We must now pass in summary review the problems connected with these factors.

Inflated Capital and Operating Costs

The excessive costs of the industry are of two sorts: (1) capital costs are inflated because more wells are drilled than are necessary and (2) all these wells entail operating costs. The costs in these two categories must be further analyzed to separate two distinguishable elements: (1) a part of the excess in costs is traceable to unused capacity; and (2) aside from excess capacity, wells are more closely spaced than is necessary for efficient drainage. It is hard to separate these two factors, since in the actual regulatory process they go hand in hand. We may start by looking at the well-spacing problem without attempting to separate them.

Well spacing is determined by rules established by regulatory agencies, subject to legislative restrictions. The incentives to drill developmental wells under the established spacing rules are provided by the depth-acreage formulas. To the extent that well factors exist in these formulas, the incentive system is distorted in the direction of drilling more wells than is necessary. For prospective fields, regulatory agencies have two options: (1) that of imposing more efficient well spacing on reservoir developers or (2) that of providing incentives to space wells more efficiently. A revision of depth schedules and proration formulas to eliminate the well factor and widen the spacing would contribute to cost reduction on new oil. There are currently strong pressures at work within the industry which make it probable that cost-saving measures of this sort will be more actively promoted by regulatory agencies and by many producers. We have already seen a substantial movement in several states to reduce or eliminate the well factor in proration formulas. Some industry representatives claim that virtually all incentives for over-drilling have been eliminated in the 1965 Texas Yardstick.

Perhaps the fatal weakness of the best as well as the worst depth-acreage schedules is that they are outdated the day after they are put into effect. For example, the apparent rationale for the 1965 Texas Yardstick is a profit one. Given today's drilling and completion costs at various depths, given today's oil prices, given today's discount rates, and given today's degree of market-demand restriction, it presumably indicates the allowables that will enable operators to earn, on the average, a reasonable profit. While it is true that none of the variables listed above change rapidly, they do change, so that a yardstick which is reasonable from an average profit standpoint today becomes a poorer and poorer approximation of an economically rational schedule tomorrow.

Thus, the rationale of depth-acreage proration schedules needs to be carefully rethought by the industry and the regulatory authorities. It is obvious that these schedules are not inducing attainable efficiency in field development and production, although the "new pool" schedule in Louisiana and the 1965 Yardstick in Texas are great improvements over the older schedules. Clearly the industry faces a difficult problem today with the imposition of heavy market-demand restrictions and, as we have noted, this compounds the problem of revising depth-acreage schedules. An alternative is to replace these schedules with something like MER regulations. The major complaint against an MER basis for allowables is that in times of severe restriction, fields with high MER's get the lion's share of allowed production and low-MER fields become unprofitable. Resort to yardsticks in times of stress is merely a scramble by low-MER producers to stay alive. If our interpretation is correct, it means that a more rational basis for allocation, such as the MER (and there may be other devices as good as or better than MER's), awaits a serious move to resolve the overcapacity problem.

However, to the extent that there are well factors still operative in the depth-acreage schedules, and they appear to exist in all the schedules we have examined, although in varying degrees, these schedules themselves will continue to contribute to the overcapacity and overinvestment problems by inducing overdrilling in many fields.

The degree of cost reduction possible through wider well spacing would still fall far short of meeting the criteria of economic efficiency since, under the existing situation, *any* new producing well in a market-demand state is, at once or after a grace period, cut back to its pro-rata share of the limited market demand. The amount that can be allocated to new wells is severely limited by two factors: (1) the large share of the market assigned to unallocated or exempt wells and (2) the share of the market assigned to allocated wells under preexisting proration formulas and spacing rules. These shares appear to be regarded as vested interests of the owners, thereby greatly limiting the amount available for assignment to

new wells on wider spacing patterns. This limitation on the allowables to new wells is a considerable deterrent upon the cost-reducing
effect of wider spacing. As long as the claims to allowables by old
producers are considered most important, the cost-reducing effects
of wider spacing will fall short of cost minimization for new production, since many new wells produce only at a low fraction of
capacity.

This fact illustrates the way in which the dead hand of history
greatly inhibits the pursuit of the goal of economic efficiency. Nevertheless, a substantial contribution to cost reduction is possible
through reforms in the proration formulas, spacing rules and depth-
acreage schedules, and these could be much extended if the forces
at work were also to lend impetus to unitization.

Excess Capacity

The heart of the cost-reduction problem, with respect to new oil,
lies in the existing state of overcapacity. This, as we saw in Chapter
5, is a long-range problem about which not very much can be done
very fast, given the present institutional environment and attitudes.
In fact, without changes in institutions and attitudes, we suspect
that the elimination of overcapacity, if this ever occurs, may well be
short-lived, since there are built-in incentives in the regulations to
generate overcapacity. The industry may have to live with excess
capacity, but can at the same time consider what might be done to
reduce it. Any solution involves the interrelationships of four elements: (1) the rate of abandonments, (2) the production-decline
rate of existing wells, (3) the rate of additions to new capacity, and
(4) the rate of secular increase in demand. If (1) plus (2) plus (4)
were certain to rapidly outrun (3), it would be possible to sit back
and let nature take its course. This is what many people hope for,
but it cannot be counted upon. For one thing, the very forces that
make the industry more profitable, i.e., decreased restrictions on
prorated wells, at the same time provide the incentives for individual
operators to go out and drill more wells to get the increased allowable. As soon as enough of these wells seeking to share an expanded
market are brought into production, capacity grows and the need for
greater restriction once more arises. This is what one would expect
to happen, since we have a system in which entry or expansion in
the industry is free of restriction, yet prices are maintained at a

relatively stable level and at a level which is, we judge, sufficiently high to return a "normal" profit on wells that often have severe restrictions on output. If allowables are raised significantly, given this price stability, profits become supranormal and are a great inducement to those with funds to invest, to put these funds into new wells.[4] Everyone who can get a producing well can have a share of the market at the going price. If this reasoning is correct, and the history of the industry gives much credence to it, then overcapacity may well be the "normal" situation for the industry, rather than a situation in which capacity is roughly equal to demand.

Related to this is the fact that the rate of growth of market demand is made problematical by the unpredictable impact of alternative sources of energy, the prospects for which are greatly improved by the relatively high marginal costs and high prices of crude oil.

The entire problem is complicated even further by the frequent and recurring remonstrances from state and federal officials and industry leaders alike urging that the oil industry exert itself to find more oil to assure the nation's proved reserves position. Thus, it may be that the official but unwritten national policy for the industry is to maintain greater reserves than are currently reported as proved. But here we run into the dilemma of how to get substantial additions to reserves while at the same time reducing excess producing capacity and the "waste" therefrom. Clearly, under the present system the achievement of both these goals simultaneously has little hope for success. Without some changes, its seems unlikely that an adequate rate of reserves discovery can be maintained without additional investment in unused capacity.

The problem is by no means insoluble, but any workable solution might have to be drastic and require substantial changes in regulatory practices and legislation. The barrier to introducing a workable solution quickly is that it would entail what would no doubt be regarded as an intolerable impairment of property rights and capital values.

For new fields it might be feasible to shut in newly discovered reservoirs after minimal drilling to establish the reserves, and re-

---

[4] The federal income-tax treatment of income from and expenses for oil drilling and production may tend to magnify the incentives for new drilling. See A. E. Kahn, "The Depletion Allowance in the Context of Cartelization," *American Economic Review*, Vol. 54 (June, 1964).

ward the investment by payments for "nonproduction," developing and producing the new reserves as needed.[5] The difficulties of determining "rewards" and the problem of who is to pay them are obvious barriers. For an industry in which compulsory unitization is suspect, in part because of a feeling of unwillingness to give this much power to a regulatory agency, it is doubtful that much enthusiasm could be generated for such a plan. Another possibility along these lines would be to permit exploration and development to occur but only on very wide spacing. Infill drilling would be prohibited. There are obvious problems in this sort of proposal also. Presumably these types of solutions could gradually reduce capacity while keeping reserves up so that eventually development could proceed with minimal restrictions on production. Proposals such as these are not very appealing to economists since they would continue the uneconomic practice of allowing inefficient reservoirs to produce while restricting efficient ones. This has the merit, however, of avoiding the investment, or at least most of it, until it is needed.

Another type of proposal has been made off and on by independent producing groups. They have advocated a closing of federal offshore lands to leasing, or at least a lengthening of the federal leases that require drilling within 5 years in order to retain the lease. The cessation of federal offshore leasing would no doubt reduce the additions to producing capacity but would also reduce additions to new reserves. Although no data are available on offshore producing capacity, we suspect that the barrels of reserves per barrel of producing capacity in offshore fields are among the highest in the country and thus are the type that could be most economically produced.[6]

Another approach to the problem, either as an alternative or as a complementary measure, would be the deliberate reduction of producing capacity in old fields. Such a reduction would require the abandonment of large numbers of wells that are superfluous to the efficient recovery of known reserves. In fact, the closing of these wells would not only diminish producing capacity but would also *increase* the percentage of oil recoverable in these fields by reducing

---

[5] Such a proposal has some rough parallels with the "soil bank" in the U.S. farm program.

[6] See memorandum to the executive committee of the IOCC from Lawrence R. Alley, "Report on Offshore Production," December 12, 1964, for a discussion of offshore production and state and federal leasing activities in Louisiana.

operating costs and thus would increase recoverable reserves. To do this and at the same time preserve the equities of property interests concerned would probably require compulsory unitization plus a withdrawal of the preferential treatment accorded to special categories of exempt wells in most market-demand states and a phasing out of much production from this source. The avenues to a much more economical structure of production are available and they need not sacrifice basic property rights, though they would necessarily cut across many vested interests.

Another alternative, related to the one immediately above, would be to determine systematically the amount of production needed to keep exempt wells operating. There is undoubtedly more production in these categories than is necessary. The restriction of output from some exempt wells would allow more production to be allocated to prorated wells. This does not reduce excess capacity, but it would help to rectify the currently perverse practice of cutting back what appear to be the lowest-cost wells. This might be a step in the direction of ridding the system of wells that cannot survive except with some form of subsidy. We shall return to this point in a moment.

Taking a fairly long-run view, it is likely that the secular rise in demand will continue and that this rise will somewhat reduce excess capacity, at least periodically. The basic question remains: Will increased allowables and profits generated by growth in demand induce additional investment in new productive capacity so that, while the industry is continually striving to reduce the excess, the excess never disappears? This, more than anything else perhaps, bothers economists, and this is the reason that some economists urge a change in the system that would allow price flexibility. Under flexible prices, it is argued, capacity is increased or decreased as demand dictates because prices will be low with overcapacity and will rise as capacity falls short of meeting demand. Incentives rise and fall with needs. We shall have occasion to look at this aspect again.

Ultimate Recovery

A thorough-going effort to place the industry on an economically efficient basis would entail the introduction of unit operation in the case of most new reservoirs and, to the extent feasible, in old reservoirs also. While state commissions have considerable power to stimulate wider well spacing, they are much more limited in their power to force the pace of unitization. In some states they appear to have been relatively indifferent to the subject. This may have been due in part to lack of statutory authority, but they probably have also entertained the antiunitization philosophy that has characterized much of the industry. It should be added that the Louisiana commission has been an exception to this generalization. It worked diligently for the passage of the compulsory-unitization statute in that state and perhaps pushed as hard as was politically feasible to encourage unit operations. Even Texas may be veering from its course of lukewarm support of unitization. The events in the Fairway Field described in Chapter 6 seem to point in this direction.

The obstacles to development on the unit plan were reviewed in Chapter 4 of this study. While these are still formidable, present cost pressures seem to be stimulating a much more active interest in the possibilities. The advantages are the same as they have always been: the saving or development expense by drilling fewer wells, the saving of operating expense because of fewer wells, enlarged ultimate recovery due to efficient use of the driving mechanism, and early introduction of pressure-maintenance measures. It is not at all clear that the industry is anywhere near the point where its predominant weight will be thrown to support effective use of state powers to advance unitization.

Another of the barriers to acceptance of unit operations is found in the production-allocation systems themselves. If an undeveloped field is being considered for unitization, the most crucial decision that the regulatory agency makes that will influence the choice of whether or not to unitize is the amount of pool allowable. Should the pool allowable be based on the yardstick, on the MER, on "recoverable" reserves, or on some other basis? Thus, we find ourselves back to the allocation system, and since the two are inextricably connected, back to the well-spacing problem also. Under unit operations, the allowed production will be one factor in determining the spacing of wells in the unit. These complicated interrelationships make us skeptical of those statements that existing opportunities for unit operations offer little in the way of higher profits. Profits depend on revenues and costs, which in turn are heavily dependent on allowables and well spacing.

Apart from pressures originating within the industry favorable to the extension of unit operation, it is conceivable that state governments, seriously reviewing their long-run interests, might conclude that they were dissipating their heritage by not throwing their weight behind unit operation. The state of Texas, for example, is going to be here a long time and it might appear shortsighted to undermine its future economic base by sterilizing a substantial portion of what might be its economically recoverable reserves. It is a nice question whether a state is capable of taking the long view when most of the political pressures come from those who are interested in making money now from their separate properties. Taking the long-run view about ultimate recovery is made more difficult by uncertainty about what will happen in the long run. In particular, there is no clear view of the extent to which alternative sources of energy will preempt the market. The time does not appear to have arrived, however, when responsible state governments can discount the importance of their potential reserves of crude oil.

It may be argued that an initially inefficient plan of reservoir development is not a throwing away of reserves, because the oil is still there and available for recovery by secondary methods. This is a partial truth that obscures the facts (1) that much oil is made physically unrecoverable by inefficient original methods, (2) that much is made economically unrecoverable, (3) that what is recovered entails a higher cost than would have been necessary under efficient original development, and (4) that the higher cost may induce earlier abandonment and thus reduce the amount of oil that could have been economically recovered.

The Bias toward High-Cost Oil

All the factors to which we have called attention cause the costs of oil to be unnecessarily high. Overinvestment and overcapacity are the most general causes. Cutting across these factors is another deliberately fostered by regulatory agencies and statutes: that of getting as much as possible of the allowed oil production from the highest-cost wells. This result is inherent in two features of the market-demand proration system: (1) the unlimited production or "bonuses" allowed to the exempt wells and (2) the way in which proration systems favor production from the less-productive wells. The oil that is *not* produced is invariably the oil that could be produced at the lowest marginal cost. This phenomenon is merely a secondary feature of a system that permits far more wells to be drilled than are necessary for efficient drainage and protects their profitability. It is, however, a popular feature, since there are far more poor wells than rich ones, and the vested interests that it creates stand as the most formidable barrier to progress toward a more economically rational structure of production. Progress in that direction requires a drastic reduction in the well population. Rapid progress would require the deliberate extinction of masses of existing wells, a task that no state agency has cared, or dared, to propose. Slower progress could be made by permitting existing wells to eke out their existence on present lines and eventually die under the present system, but putting operators on notice that for wells completed after some date, the special treatment will no longer apply. Failing such measures, the present regulatory rules contain the danger of slowing down, if not actually halting, the growth of the industry by undermining the incentives to search for new reserves.

It is the problem of innumerable small wells that brings into sharp focus the conflicting purposes of regulation: maintaining, on the one side, the access of individual owners to underlying oil and protecting their interests when once established and, on the other, eliminating waste by enforcing more economical methods of reservoir development. Wider well spacing through cooperative drilling units is a partial method of bridging this chasm of conflicting purposes, but it goes only part of the way and can act only slowly to introduce more efficient organization because of the mass of established interests to be protected and adjusted.

HIGH-COST PRODUCTION AND SUPPORTED PRICES

One effect of market-demand proration is to permit prices to be maintained at a relatively high level or, to put the matter differently, to prevent the prevailing overcapacity from undermining the price structure. This level of supported prices has two main consequences. First, it permits high-cost producers to survive or, more accurately, establishes the margin of survival. This margin is also in part determined by the unrestricted production allowed to low-yield wells. Second, even in the face of low allowables available for new wells, the supported prices provide sufficient incentive for the drilling of new development wells—in fact, far more than are needed—and for a substantial investment in exploratory work. The further such development goes, however, the more it tends to dry up the incentives except insofar as they are kept alive by additional revenues from expanding market demand.

It is interesting to speculate upon what the consequences might be of a marked increase in the productive efficiency of the industry, lowering average and marginal costs of production. It is a basic principle of competitive organization that the level of prices follows the level of costs, and therein lies one of the primary advantages of competition to the consuming public. But, as we have seen, oil prices are not determined on competitive principles. Limited production *permits* prices to be maintained at existing levels. But it does not require it. If the large integrated companies began posting lower prices, they could still get all the oil they need, at least in the short run, because the owners of existing productive capacity would have to find an outlet. However, the lower price on the oil they produce for themselves might or might not significantly affect the profits of the integrated companies. Each company would have to balance off the gains from lower-cost crude oil at the refining stage against the lower after-tax production revenues at the producing stage. As we have noted, there are great incentives for integrated companies to shift profits from refining to production because of the special expensing and depletion provisions in the federal income tax system.

As we saw in Chapter 8, there is a good deal of obscurity about the process of price determination for crude oil. The large integrated buyers presumably follow a policy which they consider to produce the greatest net advantage. But it is doubtful that this advantage can be fully expressed in accounting terms of costs and prices. Probably the principle of not upsetting the applecart is at work. An active downward tendency of posted prices would no doubt create an uproar among independent producers and might lead to retaliatory measures by state legislatures or commissions. Probably also it would lead to an intrusive interest in their pricing practices by the antitrust division of the U.S. Department of Justice, which regards one of its functions as that of protecting the "competitive" position of independent producers. The cry of "monopoly" would resound through the industry as it does whenever the declining take by the pipelines creates "distress" oil in one area or another.

It is difficult to imagine what effect a substantial improvement in the efficiency of the industry might have upon the price of oil. So long as the system of proration to market demand remained in force, it would be possible to maintain prices to the benefit of profits. It might very well be that downward pressure on prices, corresponding to lower costs, would have to come as the result of public policies rather than as the result of competitive market behavior.

There is little use in carrying on with such speculations, since it is impossible to see what changes in market behavior would accompany marked progress in the direction of efficient producing organization. Such progress is in any case bound to be slow for reasons that we elaborated earlier. But it helps to make clear the peculiar market characteristics of the industry to see that, as long as the industry is organized along its present lines with freedom of entry and a piece of the restricted market assured to every producer under public regulation and sanction, even a very marked improvement in efficient organization would not automatically convey the benefits of lower costs to the public in the form of lower prices.

PARADOXES OF POLICY

When one surveys the whole gamut of state and federal policies affecting crude oil, a basic paradox appears: some of them are stimulative and some are restrictive. Incentives are provided to discover new reserves and to develop them into producing capacity; but, once developed, the rate at which the producing capacity may be utilized is severely restricted. Both aspects are present at both state and federal policy levels. The following propositions, though greatly oversimplifying a complicated subject, will illustrate the peculiarities of the policy situation:

(1) States undertake conservation programs to "prevent waste," but do not enforce measures to ensure maximum economical recovery. This amounts to an implicit assertion that the loss of a considerable fraction of the economically recoverable reserves is not

a matter of public interest or (and this more accurately depicts the situation) is subordinate to other overriding considerations.

(2) Some states provide incentives in their allocation systems to search for new reserves by granting special quotas to discovery wells. The new fields so discovered, however, are subject to the process of partial sterilization because (a) there is no assurance that the fields are developed in the most economically efficient manner possible and (b) even when developed efficiently, the fields are often restricted to a fraction of what they can efficiently produce.

(3) While apparently careless of the reserves they waste by inefficient development, most states attach a peculiar importance to the reserves underlying their "stripper" or marginal" wells. They favor them by allowing unrestricted or at least "bonus" production, which acts as a disincentive to exploration by reducing the allowable production of prorated wells.

(4) States provide incentives to new additions to current producing capacity, via exploratory drilling, developmental drilling, secondary-recovery projects, and the like, a substantial portion of which, when added, is immediately sterilized by market-demand restrictions.

(5) Through their proration formulas, spacing rules, and depth-acreage schedules, states, in many instances, stimulate the drilling of more wells than are needed to produce the added efficient capacity, promptly cutting them back to relatively low allowables.

(6) The federal government, by special tax provisions, adds to the cash flow of funds available for new investment, much of which when invested will be sunk in unnecessary wells.

(7) The federal government, by extending the special tax provisions to companies operating abroad, stimulates their development and adds to their cash flows, but severely restricts the admission of their product to the American market under the program of oil import quotas.

(8) The federal government, by offering its own lands and offshore preserves for exploration and development, fosters additions to the already excessive capacity that has to be partially bottled up.

This confrontation of the stimulative and restrictive aspects of policies cannot be reduced to any economic rationale. It is, indeed, little more than a restatement of the argument at earlier points in this study that the industry is organized on lines that do not correspond closely to economic criteria. Cast in terms of policy rather than of industry structure, however, the elements are seen in a different light. Once the arguments for public policy are detached from the economic base, they become diffused into not necessarily consistent arguments for a variety of ends: for preventing physical wastes, for stabilizing prices, for a philosophy of property rights, for advantages to special groups, for national security, and so on. Most fields of public policy require compromises among conflicting ends, and this industry is no exception. But oil provides a peculiarly confusing interplay. What comes strongly into view is the fact that regulation is not specifically pointed toward an economically efficient, low-cost producing organization or toward maximum economic utilization of our oil resources. Other goals appear to take precedence. To the extent that they do, there is an obligation on the part of the state and federal governments to spell out and justify these goals. This has not been done.

ASPECTS OF FEDERAL INTEREST IN CRUDE OIL

In the course of this study, which is primarily concerned with the processes of state regulation, we have made only passing reference to the interest of the federal government in matters relating to oil. Before bringing the study to a close, we must give this interest explicit, if brief, attention:

(1) As a matter of general commercial policy, the United States favors trade relations designed to expand, rather than to restrict, the volume of international trade, according to principles laid down in the General Agreement on Tariffs and Trade. Under the Mandatory Import Control Program initiated in 1959, an exception is made to this policy in the case of crude oil and refined products. The total of permitted imports is set by a formula that restricts imports to a fixed percentage of domestic production,[7] and the total is divided as quotas among refiners according to their refinery runs. This exception was made on the grounds that national security con-

---

[7] Different formulas are used for the West Coast (District V) and the rest of the country (Districts I to IV) and for residual fuel oil. For Districts I to IV the present percentage is 12.2 per cent of the domestic production of crude oil and natural gas liquids.

siderations require that the search for new oil be stimulated in order to maintain a degree of self-sufficiency adequate to meet situations of national emergency. Since at the time the program was initiated, low-cost foreign oil was entering the country at a rapidly mounting rate, the effect of the program has certainly been to give strong protection to domestic producing interests. It has thus shored up the existing structure of production, possibly saving it from being thoroughly undermined.

The federal government has shown no disposition to intervene in any direct way to effect changes in the system of state oil-conservation regulations, but federal officials have on occasion, as illustrated by the statement of the Secretary of the Interior quoted earlier, expressed some dissatisfaction. Because of the strategic position it now occupies in relation to the well-being of domestic producers, the federal government is likely to give more active attention to the affairs of the industry than it has given heretofore. In fact, such attention has been forced upon it by insistent pressure from producing interests in the effort to secure greater protection. If this attention takes an analytical turn, it is almost bound to give rise to some of the lines of thought that have appeared in the present study. It may be asked, for example, whether the oil required for national security is not more costly than it need be, or whether the stimulus to exploration through import control may not be offset by factors in the regulatory system that tend to weaken the incentives to exploration. It might well happen that the federal interest in national security could become one of the forces propelling the industry in the direction of more efficient organization.

(2) There is a second aspect of the importance of oil for national security. In the existing state of international tensions, and with the possibility of outright war, it is important to the United States to maintain friendly relations with the countries possessing great oil reserves, as a means of keeping those reserves outside the political control of potential enemies. This has a restraining effect upon the degree to which import restriction is likely to be invoked. The quickest way to lose friends among oil-producing countries would be to exclude them from the American market or deny them any part of its expansion.

(3) Further, in the area of national security, one finds the state regulatory officials defending state conservation regulations on the grounds that such regulations are essential for the maintenance of

our national defense. The usual arguments are highly suspect.[8] Regardless of this, the federal government has no ongoing program to coordinate the efforts of state conservation agencies so as to achieve some stated national-security objectives. The state governments frame their argument in such a way as to deny any reason for federal intervention in oil regulatory matters except import restriction. There is without doubt a relevant question of national security with respect to the domestic oil industry. The concept, as applied to the industry, is, however, so vague and formless at the present time that it confuses rather than clarifies national or state oil policy.

If an important aspect of national policy is to maintain standby producing capacity, and if state conservation regulations are intended as one aspect of regulation to implement this goal, there should be a clear statement to this effect. Given that there is some amount of desired excess capacity, the critical question from a general public-interest standpoint is how can we minimize the cost of this capacity, which, in effect, is a national defense cost.[9] The question of whether or not there should be standby producing capacity encouraged in part by conservation regulation is important because the manner in which economic efficiency is achieved may differ greatly under conditions of no desired standby capacity from a situation in which the excess capacity is explicitly called for.

(4) Apart from considerations of national security, the federal government necessarily has a long-run interest in the future availability of adequate sources of low-cost energy. At present, a large major fraction of energy is supplied by petroleum, and of this a major fraction by crude oil. Looking to the future, when the exponential rate of economic growth will call for vastly larger supplies of energy, there is a national interest both in the discovery and

---

[8] Former Texas Railroad Commissioner, William J. Murray, has stated: "From the standpoint of public interest, *the most important reason for market demand proration* is that it permits the development of reserve producing capacity for the nation, which is vital to national security.

... If we do not practice market demand proration, the nation will at all times produce all the oil it can, and there will thus be no reserve for exceptional consumer demand or for military emergency." [Emphasis added.] "Market Demand Proration," in Wallace F. Lovejoy and I. James Pikl (eds.), *Essays on Petroleum Conservation Regulation* (Dallas: Southern Methodist University Press, 1960), pp. 68–74.

[9] We might add that there are alternative means of providing standby capacity. The comparative costs of these alternatives have never been examined carefully.

efficient development of crude-oil reserves and in the development
of alternative sources of energy. While the federal government has
never defined its policy interest in any comprehensive way, it has
intervened in various ways, as in percentage-depletion tax allowances, import control, government-sponsored research on shale oil,
and other directions. It has not, however, intervened directly, as
has at times been urged, to force upon states the use of methods of
reservoir development conducive to lower costs and higher levels of
ultimate recovery of crude oil. It would not be surprising, however,
if in due course it developed a more active interest in the possibilities. This might very well happen when the methods of refining
shale oil bring it a little closer as a competitor of crude oil. The
higher the price at which crude oil continues to be sold, the closer
is the period when shale oil can enter the market. Because so much
of the shale is located on federal public lands, the government is
in a strong position to force the competitive pace. This might be
the route by which the crude-oil industry could be propelled toward
more efficient organization. It might, however, be thought a more
rational method of developing energy sources to delay the vast investment necessary to make shale oil into a large-scale competitor
while more fully and economically developing the potential reserves
of crude oil. All these matters are veiled in uncertainties, and no
one can reasonably predict either the course of future competition
between alternative sources of energy, or the way in which federal
policies may affect it. Nevertheless, it is easy to imagine that the
competitive threat of alternative fuels might be the catalytic agent
that would force the crude-oil industry to accelerate greatly its
progress toward a more efficient state of organization.

(5) A further aspect of the federal interest in oil is represented
by the activities of the U.S. Department of Justice. Under its general
antitrust function and in the surveillance required by the law approving the Interstate Oil Compact, the department watches the
market practices of the industry. This surveillance is mainly directed
toward the market behavior of the large integrated companies, with
special reference to their control over pipelines, and it appears to be
regarded as a means of safeguarding the competitive position of
independent producers and refiners. In following this line, its influence is on the side of keeping the existing organization of the industry intact. After reading its reports, especially those of the Attorney General on the IOCC, we find ourselves in a state of some

confusion about its theory of "maintaining competition." It appears to say that the preservation of a large number of independent operators makes an industry "more competitive." One would have thought the matter could be described better the other way round: that many producers exist only because they are insulated from competitive pressures by public regulation. We have no basis for criticizing the department's view that the behavior of some pipeline owners in some areas is deplorable, but we doubt that the department has found a suitable way of translating the facts of detailed situations into a theory of competitive behavior in the industry. The whole system of conservation regulation is designed to prevent market competition. The department is perhaps somewhat addicted to an attitude widely held by independent producers who like to praise the industry for embodying the "American principle of free competitive enterprise." So far as one can assign a meaning, this seems to mean that a great many people are free to drill a great many wells at their own risk, which is true enough.

Because the industry is regulated under the authority of sovereign states, no attempt is made to apply to it the usual principle of antitrust action, which is to force efficient organization by maintaining market competition.

(6) In connection with state systems of market-demand proration, the federal government, through the Connally (Hot Oil) Act, has the power to enforce state prohibition of the movement in interstate commerce of oil produced in violation of state laws. This places the stamp of federal approval upon state methods of conservation regulation. At the same time, the federal government retains a supervisory function in that Connally Act support can be withdrawn if states are deemed to use their powers to enhance prices unduly. The active use of the powers under the Connally Act has largely been withdrawn since 1965—not for that reason, but because the problem has largely disappeared.

(7) As the owner of federal lands and the trustee of Indian lands, the federal government has direct responsibility for prescribing rules and regulations to govern exploration, development, and production. In this connection, the market-demand proration rules of the states within which the lands are located are normally applied.

In relation to none of these aspects of public interest has the federal government found occasion since World War II to intervene in order to modify the state programs of conservation regulation for crude oil. The activities of the Federal Power Commission in regulating the price of natural gas impinge upon state regulatory processes and have some indirect effect upon crude oil by affecting the relative prices and market-demand situation of natural gas and crude oil.

PROBLEMS OF CHANGE

The problems of moving from the present highly inefficient structure of the industry to a more efficient structure depend upon two factors: (1) the speed with which the transition is to be made and (2) the extent to which the principle of efficiency is to be applied. In either case, the first step is to develop new reservoirs efficiently. There is nothing further that we can usefully say on this point. The technically trained members of the industry know much more about how to do this than we do. The only prerequisite is that regulatory agencies establish rules which require that it be done.

If this were done, and time were no object, after a generation much of the present great excess of wells would have disappeared, even under existing proration systems. Even at the end of such a process, however, a state of considerable inefficiency might still exist if, though otherwise efficiently developed, reservoirs in the aggregate embodied a large amount of excess producing capacity. A state of efficiency can only be said to exist if the wells in existence are able to produce at a high percentage of capacity. A part of the problem, therefore, is to find the means by which the discovery and development of new reserves do not at the same time generate an excess of capacity. The secular growth of demand could conceivably prevent the problem from arising, but this cannot be counted on.

If it is considered important to accelerate the search for new reserves, the problem is that of modifying the existing proration system to provide additional stimulation. The way to stimulate exploration is the promise of generous production from new sources discovered. To be able to do this might require the development of a dual system of allocation—one applicable to old wells, the other to new reservoirs. New reservoirs, for example, might be placed on an MER or a reserves basis and allowed production at a relatively generous percentage of MER or reserves. Old wells could be retained on the present basis, their allowables or exempt production being limited to the extent necessary to accommodate the allocation to new reservoirs. In this case, the status of exempt wells would come under severe scrutiny and a policy of deliberately hastening their disappearance by purchase and closure might be called for.

A rational oil policy has to be oriented toward the future. Looking some distance ahead, a great deal more energy will be required than now. Of this, it may be reasonably supposed that oil, if developed along efficient lines, will remain the major constituent for at least some decades. The question for policy is how to assure that it will be forthcoming. Much is likely to be available from burgeoning foreign sources. If, however, valid reasons are found for desiring a substantial major fraction of it to be obtained from domestic sources, the policy problem is to assure the investment and enterprise that will bring this result. No one knows how much new oil can be discovered by how much effort, but it is reasonable to suppose that there is still much oil to be discovered at tolerable cost. The problem is to maintain adequate incentives for finding it.

The continuance of many present regulatory rules and practices is ill-designed to fulfill that purpose. They impose upon the industry unnecessarily high costs, and penalize the rewards for new oil in order to maintain the rewards for the owners of old producing properties. The importance of changes in regulatory practice will be their effect upon the attractiveness of future investment. Any significant effect will necessarily be associated with changes operating in the direction of lower costs. Unless changes of that sort are introduced, it appears possible that the industry will drift toward stagnation, with no increase and perhaps even a decline in activity, and with the competitive intrusion of alternative sources of energy.

Even if that were not so—even if the increase of demand were sufficient to validate high costs—it is still impossible to counter the economist's argument that it makes no sense to invest much more in an industry than is necessary to effect the productive result. There are innumerable useful outlets for investment funds, and there is no point in throwing them away to no purpose.

However much the makers of policy may desire to "follow the gleam" of the economic gods, their task will not be easy nor their progress rapid. Apart from the political opposition of vested interests, they face the fact that the great bulk of existing production comes from sources that were developed under the old and relatively inefficient system. The main opportunities for efficient development lie in new sources. Some of the existing inefficiencies could also be weeded out, but the effort to do so runs into very deep-seated attitudes concerning property rights and against damaging the interests of persons who have invested in good faith under accepted rules. It also runs into the whole complex of existing contractual relations and institutional arrangements. The process of transition from a state of inefficiency to one of efficiency will therefore be indescribably difficult and necessarily slow.

Whether the effort is worth undertaking depends upon one's views on the prospects of the oil-producing industry. If one thinks that the industry is in the later stages of exhaustion, as some do, one might easily conclude that the effort is not worthwhile. Ride out to the end on present lines with a minimum of trouble. But if one prefers to think that the oil industry has a long future and an important part to play in future economic development, as we do, then the problems become real and the lines of policy important.

Index

Milton Keynes UK
Ingram Content Group UK Ltd.
UKHW031128141024
449569UK00006B/350